STOCHASTIC MODELING IN RANGE MANAGEMENT: SELECTED ESSAYS

Stochastic Modeling in Range Management: Selected Essays

Amitrajeet A. Batabyal

Nova Science Publishers, Inc.
New York

Library of Congress Cataloging-in-Publication Data
Batabyal, Amitrajeet A., 1965-
Stochastic modeling in range management : selected essays / Amitrajeet A. Batabyal.
p. cm.
ISBN 1-59033-844-8 (hardcover)
1. Range management--Mathematical models. 2. Grazing-- Mathematical models. 3. Stochastic analysis. I. Title.
SF85.B34 2003
636.08'45'015118--dc22 203017188

400 Oser Ave, Suite 1600
Hauppauge, New York 11788-3619
Tele. 631-231-7269 Fax 631-231-8175
e-mail: Novascience@earthlink.net
Web Site: http://www.novapublishers.com

Printed in the United States of America

To Swapna and Sanjana

CONTENTS

ACKNOWLEDGMENTS

Of the twelve essays in this book, six have appeared previously in various journals. As such, I would like to take this opportunity to express my appreciation to four publishers for granting me permission to reprint these previously published essays. "On the Optimal Number of Paddocks in Short Duration Grazing," and "An Upper Bound for the Transient Resilience of a Rangeland that is Managed Using Cell Grazing," originally appeared in the *Journal of Economic Research* in 6(2), 159-169, 2001, and in 7(2), 151-160, 2002, respectively, and I thank the Hanyang Economic Research Institute and the editors of the journal for their help. "The Steady State Distribution of Animals in Short Duration Grazing," and "Herding in Ranch Operations: An Analysis of a Search Problem," originally appeared in *Applied Economics Letters* in 9(11), 695-698, 2002, and in 10(3), 177-179, 2003, respectively, and I thank Taylor and Francis <http://www.tandf.co.uk> for their assistance. "Rangeland Management Under Uncertainty: A Conceptual Approach," initially appeared in the *Journal of Range Management* in 55(1), 12-15, 2002, and I thank the editor, Gary Frasier, for his help. Finally, "On the Choice Between the Stocking Rate and Time in Range Management," originally appeared in *Environmental and Resource Economics* in 20(3), 211-223, 2001, and I thank Kluwer Academic Publishers for their kind permission.

This book would not have seen the light of day without the assistance of two groups of individuals. The first group consists of Basudeb Biswas, E. Bruce Godfrey, and Seung Jick Yoo, and I thank them for allowing me to include our co-authored work in this book. The second group consists of Cass Shellman (at RIT) and Ruby Vasquez (at USU) and I thank these ladies for all kinds of assistance with this book. In addition, I thank the Gosnell endowment at RIT for much needed financial support.

Life in general and research in particular would be significantly less interesting without the enjoyable distractions provided by my daughter Sanjana and my wife Swapna. To both these women, a big Thank You. Finally, although my father was aware of this book project that I had embarked on, he did not live to see the outcome. Nevertheless, I am optimistic that one of my staunchest supporters would have been happy with the outcome.

Amitrajeet A. Batabyal
Rochester, NY
May 2003

Chapter 1

INTRODUCTION

ABSTRACT

This book has one fundamental objective and that is to show how the theory of stochastic processes and the techniques of stochastic modeling can be used to effectively model rangelands as jointly determined ecological-economic systems, and to use stochastic rangeland models to explain various aspects of the general task of range management. To this end, this chapter first provides a brief introduction to the task of range management and then it describes the contributions of each of the book's eleven analytical chapters.

Keywords: Ecological-Economic System, Range Management, Stochastic Modeling, Theory

1. INTRODUCTION

What are rangelands? All parts of the world "that are not barren deserts, farmed, or covered by bare soil, rock, ice, or concrete can be classified as rangelands" (Holechek *et al.*, 2001, p. 1). Rangelands have been used by humans for consumptive and non-consumptive purposes for hundreds of years. Even so, at least in North America, it is only in the early 1900s that researchers first began to formally study problems associated with the management of rangelands. In this context, range management refers to "the manipulation of rangeland components to obtain the optimum combination of goods and services for society on a sustained basis" (Holechek *et al.*, 2001, p. 5). The discipline of range management developed from the early work of agronomists, general ecologists, and plant ecologists at the turn of the 20th century, largely, at the University of Nebraska.

Since then, this discipline has evolved fairly rapidly. In particular, the first range management textbook was published in 1943,[1] and the Society for Range Management (SRM) — the most prominent professional organization in range management — was founded in 1948. In the five decades since the founding of the SRM, there have been numerous advances in the field of range management. As Hoveland (2000) and Owensby (2000) have noted, these advances include the definition of apposite stocking rate and animal performance relationships, the refinement of prescribed burning techniques, the development

---

[1] See Stoddart and Smith (1943).

of suitable plant varieties and seeding methods for reseeding, the renovation of pastures with legumes, and the recycling of agricultural wastes in forage production.

Although these advances have certainly increased our understanding of aspects of the general task of range management, surprisingly, this progress has been achieved in a largely atheoretical context. In other words, although empirical and case study based analyses of range management abound, there are very few *theoretical* studies of range management. Moreover, even though rangelands are jointly determined *ecological-economic systems*,[2] a large part of the existing literature in this field has viewed rangelands as fundamentally ecological and not as ecological-economic systems.

The few existing studies of range management that are theoretical in nature are often the result of research by economists. For the most part, these studies view ranchers as conventional neoclassical cost or profit optimizers and hence the task of range management frequently becomes a problem in constrained optimization.[3] We stress that there is nothing inherently disquieting about the use of the constrained optimization approach. Our point here is simply that although rangelands ought to be viewed as integrated or jointly determined ecological-economic systems, until very recently, the existing literature — both theoretical and atheoretical — has not viewed rangelands in this way.

Now, a crucial point about range management is that it involves decision making under *uncertainty*. At the conceptual level, this basic point has been recognized by Heady and Child (1994), by Holechek *et al.* (2001), and by many others. Even so, once again, it is fair to say that there are very few studies of range management that explicitly model this uncertainty and the impact that this uncertainty has on decision making by range managers.

Our discussion thus far yields three conclusions. First, there are very few theoretical studies of range management. Second, generally speaking, the extant literature on range management has not thought of rangelands as jointly determined ecological-economic systems. In fact, this literature has regarded rangelands as either ecological systems or as economic systems. Finally, although in principle researchers have recognized the point that effective range management involves decision making under uncertainty, there is a distinct lacuna in the existing literature as far as the modeling of uncertainty is concerned.

Given this state of affairs, the purpose of this book is to address the above three points thoroughly. In particular, we demonstrate how the theory of stochastic processes and the techniques of stochastic modeling can be used to construct and analyze theoretical models of jointly determined rangelands and thereby shed light on the task of range management. It is important to note that specialized grazing systems involving the *rotation* of animals between pastures "have been a major focus of range researchers and managers since the 1950s" (Holechek *et al.*, 2001, p. 247). In particular, this rotation phenomenon is an attribute of several specialized grazing systems such as deferred rotation grazing, the Merrill three-herd and four-pasture system, rest rotation grazing, and short duration grazing. Of all the specialized grazing systems that have been and continue to be used by range managers, short duration grazing is arguably the most prominent. Therefore, in many of this book's chapters, we acknowledge this fact by studying stochastic models of short duration grazing. We now summarize the intellectual contributions of the individual chapters in this book.

2 For more on this, see Perrings and Walker (1995) and Batabyal (2002).
3 For instance, see Karp and Pope (1984) and McCluskey and Rausser (1999).

2. THE INDIVIDUAL ESSAYS

2.1. Four Outstanding Issues

Spatial and temporal controls have been commonly employed to manage dynamic and stochastic ecological-economic systems such as rangelands. This notwithstanding, we still know very little about their *theoretical* properties. As pointed out in section 1, this is primarily because there are very few theoretical studies that provide an integrated ecological-economic perspective on how spatial and temporal controls influence the ecology and the economics of rangelands. Now, economists have studied the subject of range management in considerable detail.[4] Even so, these analyses have paid scant attention to the *ecological* aspects of the management problem or to the use of space and time as control variables. Similarly, although there are many ecological studies of range management,[5] these studies have rarely analyzed spatial and temporal controls and the *economic* effects of such controls.

As such, within the context of short duration grazing, chapter 2 identifies and describes a theoretical research agenda involving four outstanding issues in range management. The first issue of interest concerns the optimal number of paddocks and the specific question of interest is this: What is the optimal number of paddocks for rangelands that are managed with short duration grazing? This chapter suggests that queuing theory be used to answer this question. Why? This is because queuing theory can be used to derive a criterion function that can in turn be optimized to give us the optimal number of paddocks.

The second issue that chapter 2 focuses on concerns the temporal control aspect of range management. Specifically, we would like to know how a range manager ought to set the temporal control optimally for a rangeland that is managed with short duration grazing. This chapter suggests the use of a renewal theoretic framework on the ground that such a framework can be used to derive an objective function for a range manager that is a function of the temporal control. Once this has been done, chapter 2 points out that the optimal temporal control can be chosen to optimize this objective function.

We have already noted that effective range management involves decision making under uncertainty. The third issue of interest in chapter 2 involves an investigation of a particular stochastic property for our prototypical specialized grazing system, namely, short duration grazing. The specific question of interest concerns the determination of the steady state distribution of animals for rangelands that are managed using short duration grazing, and, on grounds of generality, chapter 2 suggests the use of an alternating renewal process to address this question.

The final issue addressed in chapter 2 concerns the notion of an irreversibility. In particular, this chapter first notes that it is possible for a rangeland to exist in irreversible or crisis states. The chapter then argues that in such a setting, a range manager's goal should be to maximize the amount of time the rangeland under study spends in the good set of states. Specifically, chapter 2 suggests that we study the properties of the following sensible approach: First, identify the crisis state or states. Although it is generally not possible to do so exactly, on the basis of prior experience and obtainable scientific information, range

4 For examples, see Karp and Pope (1984), Huffaker *et al.* (1989), and Torrell *et al.* (1991).

5 Such as Graetz (1986), Gutman *et al.* (1999), and Rehman *et al.* (1999).

managers are often able to specify a safe minimum standard,[6] and all the states of the rangeland below the safe minimum standard can be thought of as crisis states. Next, put in place a well designed plan of action. In the context of a specific rangeland, it is possible that such a plan of action will involve the use of a specialized grazing system such as short duration grazing.

2.2. The Optimal Number of Paddocks

Chapters 3 through 6 follow the research agenda suggested in chapter 2 and these chapters also provide a more detailed analysis of the four outstanding issues mentioned in chapter 2. In particular, chapter 3 begins by pointing out that even though spatial and temporal controls are used in short duration grazing, we still know very little about how a manager ought to go about the task of selecting the optimal number of paddocks for a rangeland that is managed with short duration grazing. Therefore, chapter 3 discusses how the optimal number of paddocks ought to be chosen for rangelands that are managed using short duration grazing.

The contribution of this chapter is that it shows how queuing theory can be used to shed light on the above question. In particular, chapter 3 shows that three steps that are specific to short duration grazing can be modeled nicely by using a specific kind of closed network queue known as the cyclic queue.[7] In a cyclic queue, neither customers (the animals) nor servers (the paddocks) are permitted to enter or leave the queuing system (the N paddock short duration grazing system). Further, the animals move from one paddock to the next in a cyclical fashion so that the output of, say, paddock i is the input to paddock $i+1$ and, in particular, the output of the last paddock N is the input to the first paddock 1.

Following this demonstration, chapter 3 computes the stationary probabilities for the N paddock short duration grazing system under study. It then sets up an expected profit maximization problem in which the natural number N is an argument of the expected profit function. Finally, this chapter demonstrates that the optimal number of paddocks is the solution to this expected profit maximization problem.

2.3. The Optimal Temporal Control

Chapter 4 analyzes the second outstanding issue mentioned in chapter 2. Now, once the optimal number of paddocks has been determined, the next salient step in short duration grazing involves ascertaining the optimal length of time during which a manager should permit his or her animals to graze a particular paddock. To this end, chapter 4 begins by pointing out that even though temporal controls are a non-trivial part of the use of short duration grazing as a management tool, we still know very little about the theoretical properties of optimal temporal controls in short duration grazing.

6 See Campbell and Stafford Smith (2000) for additional details on practical range management under uncertainty. See Farmer and Randall (1998) for a general discussion of the safe minimum standard in the presence of potential irreversibilities.

7 See Gross and Harris (1985) and Ross (2000) for more on cyclic queues.

Given this state of affairs, how should a manager select the temporal control optimally in short duration grazing? This question is analyzed in chapter 4. The stochastic modeling that is undertaken in this chapter involves the use of renewal theory — see Ross (1996; 2000) — and what drives the theoretical analysis of this chapter is the idea that the range manager will stop grazing in a particular paddock and move his or her herd of animals to the next paddock either when T time periods have passed, or when a random and deleterious environmental event curtails the operative lifetime of this paddock, thereby leading to a breakdown of this paddock as far as subsequent grazing is concerned.[8]

Because there are two possible contingencies that will terminate grazing on a paddock, chapter 4 first computes the expected net cost arising from these contingencies in a grazing cycle and then it uses the renewal-reward theorem — see Ross (1996, p. 133) — to calculate the range manager's long run expected net cost. This long run expected net cost function is constructed in a way so that the key variable of interest, i.e., the temporal control, appears as an argument of this function.

Once this has been done, chapter 4 shows that the optimal temporal control minimizes the range manager's long run expected net cost function. The general optimization problem that leads to this conclusion is unwieldy. Consequently, this chapter uses the uniform and the exponential distribution functions to show how one can solve for the optimal temporal control in specific instances.

2.4. The Steady State Distribution of Animals

The third outstanding issue that chapter 2 refers to is the steady state distribution of animals in short duration grazing and this issue is also the subject of chapter 5. This chapter begins by echoing a general claim of this book that despite the fundamental role that uncertainty plays in effective range management in general and in short duration grazing in particular, there are very few studies of short duration grazing in a stochastic framework.

As such, chapter 5 uses a particular kind of stochastic process known as the alternating renewal process to delineate a rangeland that is managed using short duration grazing. The appeal of the alternating renewal process approach is twofold. First, when this stochastic process is used, it is not necessary to use specific functional forms for the distribution functions to model the random amount of time that the manager's animals spend in the various paddocks of a rangeland. Second, it is possible to model the amount of time spent by the animals in separate paddocks with distinct distribution functions.

Now, the amount of time spent by the manager's herd of animals in any paddock is obviously a non-negative random variable. This fact is used in chapter 5 to first compute the mean amount of time spent by the herd of animals in an arbitrary (say *ith)* paddock of the rangeland under study. Then, in the limit as time approaches infinity, the stationary probability that the herd of animals is grazing the *ith* paddock is given by a simple ratio of appropriate stationary probabilities. In this way, the entire steady state probability distribution function of animals for all the paddocks of the rangeland can be easily ascertained.

---

8 Examples of such injurious environmental events include unusually high rainfall in an arid paddock that results in significant topsoil loss and the appearance of blood sucking insects that stress cattle on the paddock. See Holechek *et al.* (2001, p. 421) for additional details.

2.5. Irreversibilities in Range Management

The fourth and final outstanding issue of chapter 2 is analyzed in detail in chapter 6. This chapter begins by considering a rangeland that can exist in a finite number of states. Some of these states are good but others are bad. In addition, in both these sets of states, some states are better than others. Now, what is important for our purpose is that some of these bad states may well be *irreversible*. In these irreversible or *crisis* states, the rangeland under study is so degraded that no matter how hard a manager might try, (s)he will be unable to shift the rangeland to any other state. Chapter 6 takes the position that in this situation, the task of rangeland management ought to be viewed as one in which a range manager's aim is to *minimize* the amount of time the rangeland spends in the bad, but not the irreversible, set of states.

Given this perspective and because it is always *possible* that a managed rangeland will hit a crisis state, chapter 6 uses a discrete-time Markov chain theoretic approach[9] to first describe and then analyze a slack and a strict management regime. For each of these two regimes, the key question that this chapter focuses on is the expected amount of time it takes to hit the irreversible or crisis state.

Chapter 6 shows that for both the regimes being analyzed, the expected time until crisis depends essentially on the logarithm of the state in which the manager inherits the rangeland being studied. Therefore, the condition in which a manager inherits a rangeland is a significant determinant of the amount of time it takes to get to the crisis state. Further, in comparison with the slack management regime, it takes longer to hit the crisis state with a strict management regime. The way in which these stochastic models might be operationalized is illustrated with a numerical example that uses a routinely employed four condition class classification scheme for rangelands (Holechek *et al.*, 2001, p. 185).

2.6. A Conceptual Approach

There are very few analyses that have expressly modeled the nexuses between uncertainty, range ecology, and range management over time. Therefore, chapter 7 points out that a conceptual approach is generally needed to provide guidance for empirical analyses pertaining to policies that influence the use of rangelands. Following this observation, chapter 7 uses the theory of discrete-time Markov chains to provide this conceptual approach.

The approach developed in this chapter has three desirable features. First, this approach is consistent with the justly prominent state-and-transition model of range behavior proposed by Westoby *et al.* (1989). Second, this approach plainly accounts for the part that uncertainty plays in the dynamic behavior of managed rangelands. Finally, and perhaps most importantly, this approach demonstrates when policy has a role to play and when it does not have much of a role to play in ensuring the sustainable use of rangelands.

The discrete-time Markov chain theoretic model that chapter 7 constructs uses the commonly employed four condition class classification scheme for rangelands (Holechek *et al.*, 2001, p. 185).[10] Further, following the line of reasoning employed in chapter 6, this

[9] For more on discrete-time Markov chains, see Ross (1996; 2000).

[10] The reader will recall that this scheme was also used in chapter 6.

chapter divides the four states or condition classes of the rangeland under study into a good and a bad set of states and then it shows that the four state transition probability matrix of the rangeland under study can be split up into four smaller sub-matrices. These sub-matrices correspond to four zones as far as range policy efficacy is concerned. As this chapter shows, in two out of the four zones, there is little that policy can do to influence the behavior of the rangeland under study. In contrast, in the remaining two zones, there is a lot that a manager can do to shape the future behavior of the pertinent rangeland.

A final contribution of this chapter is the computation of the expected amount of time the rangeland spends in the good and in the bad set of states. These expectations are meaningful because of two reasons. First, they can be thought of as representing the ecological stability property known as *persistence*.[11] Second, from a conceptual standpoint, a range manager may want to think of these expectations as objective functions that (s)he might want to optimize. In this way of looking at the task of range management, policies would need to be put in place so that the persistence of the good (bad) set of states is effectively maximized (minimized).

2.7. The Choice between the Stocking Rate and Time

If a range manager is able to choose between the *stocking rate* and the length of *time* during which his or her animals graze a particular rangeland, then which control variable should (s)he choose? There is a debate in the extant range management literature about this issue. Standard textbooks such as Stoddart *et al.* (1975) and Holechek *et al.* (2001) claim that the stocking rate is more important than time. On the other hand, Savory (1988), Savory and Butterfield (1999), and other like-minded researchers claim that overgrazing has less to do with the stocking rate and more to do with the length of time that a rangeland is exposed to animals.

As chapter 8 points out, although there are numerous empirical and case study based analyses of this question, there are *no* theoretical studies. Consequently, the purpose of chapter 8 is to construct and analyze a stochastic model that sheds theoretical light on this acrimonious controversy. Following the lead of chapters 4 and 5, this chapter too uses a renewal theoretic framework to study the choice issue that we have just discussed.

Further, like chapter 4, the two principal sections of chapter 8 use renewal theory — and the renewal-reward theorem in particular — to derive the long run expected net cost ($LRENC$) of a rancher who obtains benefits and incurs costs from range operations. The working assumption here is that this rancher wishes to minimize his or her $LRENC$. In the first principal section, this rancher chooses the stocking rate optimally to minimize the $LRENC$ and in the second principal section, this same rancher chooses the length of the grazing cycle optimally to minimize the $LRENC$.

The critical question now pertains to the magnitudes of these two minimized $LRENC$ s. The stochastic modeling in chapter 8 tells us that the rancher's $LRENC$ with an optimally selected stocking rate is *lower* than his or her $LRENC$ with an optimally selected grazing cycle length. What this tells us is that if a rational rancher had to select a single choice variable from a choice set consisting of the stocking rate and the grazing cycle length (time), then this rancher would select the stocking rate over time.

---

[11] See Pimm (1991) for additional details on this stability concept.

2.8. The Upper Bound

Given the significance of *transient* behavior for the efficacious management of ecological-economic systems such as rangelands, one can ask the following question: What do we presently know about transient resilience[12] in the context of rangelands? Regrettably, the answer is very little. The economics literature appears to have ignored this issue completely (Batabyal, 2000). Even in the range management literature, there are very few analyses of rangeland resilience. Bork *et al.* (1998) and Perevolotsky *et al.* (2001) have studied, respectively, the resilience of a three-tip sagebrush range in Idaho and the resilience of the prickly burnet in eastern Mediterranean ranges. However, consistent with a central claim in this book, these two papers are empirical in nature and they do not employ an explicitly stochastic framework. In addition, neither of these two papers are concerned with short duration grazing.

Given this state of affairs, chapter 9 has two objectives. First, this chapter use a stochastic framework to model a rangeland that is managed using cell grazing. Next, this chapter calculates a tight upper bound for the transient resilience of this rangeland. For the kind of stochastic model that is used in chapter 9, it is not possible to obtain a closed-form expression for a rangeland's transient resilience. This is why this chapter computes an upper bound and not an exact expression for a rangeland's transient resilience.

Using its stochastic framework, chapter 9 first points out that for a rangeland or cell that is managed with short duration grazing, there will be a change in system control or structure when a sufficient number of paddocks are degraded so that, for all intents and purposes, the rangeland itself is degraded. In chapter 9, this sufficient number is denoted by k, where $k \le n$, and n refers to the total number of paddocks into which the rangeland itself has been divided. Transient resilience is defined to be the probability that the rangeland is not degraded. These observations are then used to compute the upper bound for the transient resilience of the rangeland under study. This bound depends on the total number of paddocks in the cell (n), the probability that a specific paddock is degraded in the course of short duration grazing (p), and the sufficient number of paddocks (k) alluded to above.

2.9. Herding in Ranch Operations

Present-day ranch operations in the United States involve two kinds of herding activities. The first kind of herding involves making sure that a rancher's animals stay in a specific pasture or allotment and that they don't wander off elsewhere. The second kind of herding involves physically moving animals from one pasture to another pasture or, alternately, to a Bureau of Land Management (BLM) allotment.

---

12 Two notions of resilience are prevalent in the ecology literature. Resilience in the sense of Holling refers to "the amount of disturbance that can be sustained [by an ecosystem] before a change in system control or structure occurs" (Holling *et al.*, 1995, p. 50). The Pimm (1984) notion of resilience focuses on stability near an equilibrium stationary state. Resistance to shocks and speed of return to the equilibrium are used to measure this notion of resilience. In contrast, the Holling notion of resilience focuses on conditions far away from an equilibrium stationary state where shocks can flip an ecosystem into another stability domain. We are using the term resilience in the sense of Holling.

When a rancher's animals are missing, this individual has to determine whether to commit costly resources and search for them. Now, as is well known, depending on the odds of finding the missing animals, a decision to search can have important financial repercussions for this rancher. Therefore, under what scenarios should a rancher search for his or her missing animals? This important question is analyzed in chapter 10.

This chapter uses the theory of Poisson processes — see Ross (1996; 2000) — to shed light on three interesting issues. First, it ascertains the rancher's expected benefit function from searching for his or her missing animals. Second, it calculates the optimal length of time for which this rancher ought to be searching for his or her missing animals. Finally, this chapter compares the merits of static and dynamic search policies. In the stochastic setting of chapter 10, a static search policy is one which involves searching for a fixed but arbitrary period of time. In contrast, a dynamic search policy permits the decision to stop searching at any time t to depend on the number of missing animals found by time t.

The analysis in chapter 10 leads to two interesting conclusions. First, it is shown that if a specific inequality holds, then it is optimal for the rancher to search for his or her animals for a positive time period. On the other hand, if this inequality doesn't hold, then it is optimal to let the missing animals go and not search at all. Second, it is demonstrated that static and dynamic search policies are equivalent and hence the rancher under consideration in this chapter is as well served by a static policy as he or she is with a dynamic search policy.

2.10. Range Condition Assessment

There are two approaches to evaluating the condition or state of a rangeland. The first approach is based on the work of Dyksterhuis (1949) and the second or site potential approach is based on the work of Humphrey (1949). Unfortunately, neither of these two approaches accounts for the fact that the spatial and the temporal behavior of most rangelands is subject to the presence of *thresholds*. A threshold indicates the boundary between two states of a rangeland in space and in time. In addition, the shift from a pre-threshold state to a post-threshold state is generally reversible only at great cost.

Despite the salience of thresholds in determining the condition of a range, there is virtually no theoretical research on range condition assessment in the presence of a threshold. Consequently, there are many significant questions in this area that confront researchers. Chapter 11 addresses two of these salient questions. To fix ideas, consider the case of a single threshold. Then, the first question addressed in chapter 11 involves a determination of the theoretical relationship between the pre-threshold and the post-threshold states of the rangeland under study. The second question involves a computation of the expected amount of time the rangeland spends in the pre-threshold and in the post-threshold states.

This chapter uses a discrete-time Markov chain theoretic model — very much like the model used in chapter 7 — to study the two questions mentioned in the previous paragraph. As far as the theoretical relationship between the pre-threshold and the post-threshold states is concerned, the analysis in this chapter shows that the rangeland under study crosses the threshold at a rate equaling the product of the stationary and the transition probabilities, added over all possible states that the rangeland can exist in. In addition, this chapter calculates the mean times that the rangeland under study spends in the pre-threshold and in the post-threshold states and then it points out that these two expectations can be regarded as a twofold measure of the condition of the rangeland.

2.11. The Role of Time

In a prominent book titled *Holistic Resource Management*, Allan Savory (1988) forcefully argued in favor of adopting, *inter alia*, a time based approach to range management. According to him and other like-minded professionals, overgrazing bears "little relationship to the number of animals but rather to the *time* plants [are] exposed to the animals" (Savory and Butterfield, 1999, p. 46).[13] Although this view has gained currency in the range management literature, there are *no* studies that provide an integrated ecological-economic perspective on how a time based management regime influences the ecology and the economics of rangelands. Consequently, very little is known about the theoretical properties of time based range management regimes.

Given this state of affairs, chapter 12 has three goals. First, this chapter uses the theory of semi-Markov processes — see Ross (1996; 2000) — to construct a stochastic ecological-economic model of a rangeland in which time restrictions are used to manage the rangeland. Second, this chapter derives two managerial objective functions that are meaningful from both an ecological and an economic standpoint. Finally, using these two objective functions, chapter 12 discusses an approach to range management in which the manager has two interrelated goals. This manager maximizes the profit from range operations and (s)he also takes steps to move the rangeland away from the least desirable state of existence.

The analysis conducted in chapter 12 leads to four noteworthy conclusions. First, unlike most economics studies on this subject, chapter 12 demonstrates that fruitful range management involves paying attention to both the ecology and to the economics of the rangeland under consideration. Second, this chapter notes that by maximizing *long run* profit functions, a range manager will also be augmenting the resilience of the good states and lowering the resilience of the bad state. Third, this chapter analyzes two different ways that can be used by a range manager to set the temporal control optimally in an ecological-economic context. Finally, this chapter identifies the probabilities that will need to be estimated in order to operationalize the stochastic models that are developed and discussed here.

3. Conclusions

A central claim of this book is that there are very few theoretical studies of range management. As such, chapters 2 through 12 in this book have provided alternate theoretical perspectives on the task of range management. Second, because the extant literature on range management has not, until very recently, thought of rangelands as jointly determined ecological-economic systems, several chapters in this book have analyzed the task of range management from an explicitly interdisciplinary perspective. Finally, we noted in section 1 that although researchers have acknowledged the point that competent range management

[13] The role of time in range management is discussed in chapter 8 as well. See section 2.7 for additional details.

involves decision making under uncertainty, there is a clear gap in the existing literature as far as the modeling of uncertainty is concerned. Therefore, chapters 2 through 12 demonstrate several ways in which the theory of stochastic processes and the techniques of stochastic modeling can be fruitfully combined to construct and analyze models of rangelands that not only account for uncertainty explicitly but also show how the presence of uncertainty alters one's perspective on what constitutes effective range management.

REFERENCES

Batabyal, A.A. 2000. Quantifying the Transient Response of Ecological-Economic Systems to Perturbations, *Environmental Impact Assessment Review*, 20, 125-133.

_____ . 2002. Necessary and Sufficient Conditions for the Equivalence of Economic and Ecological Criteria in Range Management. Forthcoming, *Journal of Economic Behavior and Organization.*

Bork, E.W., West, N.E., and Walker, J.W. 1998. Cover Components on Long-Term Seasonal Sheep Grazing Treatments in Three-Tip Sagebrush Steppe, *Journal of Range Management*, 51, 293-300.

Campbell, B.D., and Stafford Smith, D.M. 2000. A Synthesis of Recent Global Change Research on Pasture and Rangeland Production: Reduced Uncertainties and Their Management Implications, *Agriculture, Ecosystems, and Environment*, 82, 39-55.

Dyksterhuis, E.J. 1949. Condition and Management of Rangeland Based on Quantitative Ecology, *Journal of Range Management*, 2, 104-115.

Farmer, M.C., and Randall, A. 1998. The Rationality of a Safe Minimum Standard, *Land Economics*, 74, 287-302.

Graetz, R.D. 1986. A Comparative Study of Sheep Grazing in a Semi-Arid Saltbush Pasture in Two Condition Classes, *Australian Rangeland Journal*, 8, 46-56.

Gross, D., and Harris, C.M. 1985. *Fundamentals of Queuing Theory*, 2nd edition. Wiley, New York, NY.

Gutman, M., Holzer, Z., Baram, H., Noy-Meir, I., and Seligman, N.G. 1999. Heavy Stocking and Early-Season Deferment of Grazing on Mediterranean-Type Grassland, *Journal of Range Management*, 52, 590-599.

Heady, H.F., and Child, R.D. 1994. *Rangeland Ecology and Management.* Westview Press, Boulder, CO.

Holechek, J.R., Pieper, R.D., and Herbel, C.H. 2001. *Range Management*, 4th edition. Prentice Hall, Upper Saddle River, NJ.

Holling, C.S., Schindler, D.W., Walker, B.W., and Roughgarden, J. 1995. Biodiversity in the Functioning of Ecosystems: An Ecological Synthesis, in C. Perrings, K.G. Maler, C. Folke, C.S. Holling, and B.O. Jansson, (Eds.), *Biodiversity Loss*. Cambridge University Press, Cambridge, UK.

Hoveland, C.S. 2000. Achievements in Management and Utilization of Southern Grasslands, *Journal of Range Management*, 53, 17-22.

Huffaker, R.G., Wilen, J.E., and Gardner, B.D. 1989. Multiple Use Benefits on Public Rangelands: An Incentive-Based Fee System, *American Journal of Agricultural Economics*, 71, 670-678.

Humphrey, R.R. 1949. Field Comments on the Range Condition Method of Forage Survey, *Journal of Range Management*, 2, 1-10.

Karp, L.S., and Pope, A. 1984. Range Management Under Uncertainty, *American Journal of Agricultural Economics*, 66, 437-446.

McCluskey, J.J., and Rausser, G.C. 1999. Federal Grazing Reform and Avoidable Risk, *Journal of Agricultural and Resource Economics*, 24, 140-154.

Owensby, C.E. 2000. Achievements in Management and Utilization of Privately-Owned Rangelands, *Journal of Range Management*, 53, 12-16.

Perevolotsky, A., Ne'eman, G., Yonatan, R., and Henkin, Z. 2001. Resilience of Prickly Burnet to Management in East Mediterranean Rangelands, *Journal of Range Management*, 54, 561-566.

Perrings, C., and Walker, B. 1995. Biodiversity Loss and the Economics of Discontinuous Change in Semiarid Rangelands, in C. Perrings, K.G. Maler, C. Folke, C.S. Holling, and B.O. Jansson, (Eds.), *Biodiversity Loss*. Cambridge University Press, Cambridge, UK.

Pimm, S.L. 1984. The Complexity and Stability of Ecosystems, *Nature*, 307, 321-326.

_____ . 1991. *The Balance of Nature?* University of Chicago Press, Chicago, IL.

Rehman, A., MacKintosh, J.B., Warren, B.E., and Lindsay, D.R. 1999. Revegetated Saline Pastures as a Forage Reserve for Sheep: 1. Effects of Season and Grazing on Morphology and Nutritive Value of Saltbush, *Rangeland Journal*, 21, 3-12.

Ross, S.M. 1996. *Stochastic Processes*, 2nd edition. Wiley, New York, NY.

Ross, S.M. 2000. *Introduction to Probability Models*, 7th edition. Academic Press, San Diego, CA.

Savory, A. 1988. *Holistic Resource Management*. Island Press, Washington, DC.

Savory, A., and Butterfield, J. 1999. *Holistic Management*, 2nd edition. Island Press, Washington, DC.

Stoddart, L.A., and Smith, A.D. 1943. *Range Management*. McGraw-Hill, New York, NY.

Stoddart, L.A., Smith, A.D., and Box, T.W. 1975. *Range Management*, 3rd edition. McGraw Hill, New York, NY.

Torrell, L.A., Lyon, K.S., and Godfrey, E.B. 1991. Long Run versus Short Run Planning Horizons and the Rangeland Stocking Rate Decision, *American Journal of Agricultural Economics*, 73, 795-807.

Westoby, M., Walker, B., and Noy-Meir, I. 1989. Opportunistic Management for Rangelands Not at Equilibrium, *Journal of Range Management*, 42, 266-274.

Chapter 2

CONTEMPORARY THEORETICAL RESEARCH IN RANGE MANAGEMENT: FOUR OUTSTANDING ISSUES

ABSTRACT

In recent times, ecologists and economists have drawn attention to the fact that rangelands ought to be viewed and studied as jointly determined ecological-economic systems. However, because this observation has been rather recent, a number of theoretical issues in range management remain poorly understood. Consequently, the purpose of this chapter is to identify and discuss four of these outstanding issues.

Keywords: Dynamic, Ecological-Economic System, Management, Rangeland, Stochastic

1. INTRODUCTION

Spatial and temporal controls have been widely used to manage dynamic and stochastic ecological-economic systems that are jointly determined. Examples of such systems include hunting grounds, fisheries, and rangelands.[1] Weninger and Strand (1998) note that recreational and commercial hunters for most game are subject to seasonal restrictions. As well, such hunters are generally required to hunt during daylight hours. In addition to this, in almost every state in the USA, sport fishing seasons exist for a whole host of fish species. Commercial fisheries in Canada, the USA, and in western Europe are subject to a variety of spatial and temporal controls.[2] As observed by Young (1981), fisheries in the US Pacific Northwest are routinely regulated with season length restrictions (a temporal control) and with area restrictions (a spatial control). Hartwick and Olewiler (1998, p. 156) have pointed out that on the US eastern seaboard, the Atlantic surf clam fishery has been regulated by

---

1 "All areas of the world that are not barren deserts, farmed, or covered by bare soil, rock, ice, or concrete can be classified as rangelands. Therefore, rangelands include deserts and forests and all natural grasslands" (Holechek *et al.*, 2001, p. 1). In the Southern African context, our use of the term rangeland refers primarily to communal grazing lands in which the grazing animals are privately owned but the land itself is communally owned.

2 For a more detailed corroboration of this claim, see Young (1981), Karpoff (1987), Matulich *et al.* (1996), Hartwick and Olewiler (1998, pp. 152-175), and Weninger and Strand (1998).

controlling the days per week and the hours per day that fishing vessels can harvest (temporal controls), and by restricting harvests in areas where young clams reside (a spatial control).

Specialized grazing systems have been used to manage rangelands[3] in North America, in Southern Africa, and in other parts of the world at least since the 1950s.[4] The management of rangelands with these grazing systems typically involves the use of spatial and temporal controls. Spatial controls tell us which pastures can be grazed at a particular point in time and temporal controls specify the length of time during which animals may graze a particular pasture. Indeed, as Holechek *et al.* (2001) have pointed out, grazing systems such as deferred rotation grazing, the Merrill three herd and four pasture system, rest rotation grazing, and particularly short duration grazing (also called cell grazing), all have spatial and temporal control aspects to them. However, the spatial and the temporal control features of range management can be seen most clearly in the case of short duration grazing. The discussion in this and the previous paragraph tells us that as far as the management of jointly determined ecological-economic systems such as hunting grounds, fisheries, and rangelands is concerned, the use of spatial and temporal controls is standard.

In the rest of this chapter we shall focus on rangelands as our prototypical ecological-economic system,[5] and on short duration grazing[6] as the representative grazing system. We focus on short duration grazing for four reasons. First, the history of short duration grazing begins in Southern Africa. Indeed, short duration grazing as we know it today was first practiced in what is now Zimbabwe by Allan Savory and the desirable properties of this grazing system were first studied and discussed in the context of Savory's experiences in Zimbabwe (see Savory (1983) and Savory and Butterfield (1999)). Second, in addition to Zimbabwe, short duration grazing has been used in South Africa as well. As noted in Wilcox (1976), this grazing system (called management system in Wilcox (1976)) has been frequently used in South Africa along with other systems such as controlled selective grazing and non-selective grazing. Third, short duration grazing has a number of features in common with other grazing systems that are used in Southern Africa and in other parts of the world. For instance, the rotation of animals is an integral component of short duration grazing and rotation is an important concept in range management in Southern Africa as well. For instance, pointing to the desirability of rotating animals in Zimbabwian rangelands, Scoones (1996, p. 218) has noted that "deferred grazing rotations can ensure late dry season fodder reserves." Finally, a key concern of short duration grazing is to prevent rangeland degradation (see Savory and Butterfield (1999)) and this is arguably also the most important issue in the management of communal African rangelands. As Behnke and Scoones (1993, p. 2) have noted, the "preeminent management problem on communal African rangeland...[is] the control of rangeland degradation..." Largely in agreement, Abel (1993, p. 173) has observed that "the main problems of livestock development in sub-Saharan Africa...[are] over-accumulation of stock...and land degradation."

---

3 Range management "is the manipulation of rangeland components to obtain the optimum combination of goods and services for society on a sustained basis (Holechek *et al.*, 2001, p. 5).

4 For North America see Holechek *et al.* (2001), for South Africa see Wilcox (1976), for Zimbabwe see Scoones (1996), and for the rest of the world see Behnke *et al.* (1993), Holechek *et al.* (2001), and the many references in these last two sources.

5 However, we stress that our analysis applies equally well to fisheries.

6 For more on short duration grazing, see Savory (1983) and Savory and Butterfield (1999).

The view that rangelands ought to be studied as a jointly determined ecological-economic system (see Perrings and Walker (1995)) is very recent. Consequently, even though spatial and temporal controls have been and are used frequently to manage rangelands, we still know very little about their *theoretical* properties. This is largely because there are *no theoretical analyses* that provide an integrated ecological-economic perspective on how spatial and temporal controls affect the ecology and the economics of rangelands. Economists have studied the subject of range management in considerable detail.[7] However, these studies have rarely paid any attention to the *ecological* aspects of the management problem or to the use of space and time as control variables. Similarly, although there are many ecological studies of range management,[8] these studies have rarely analyzed spatial and temporal controls and the *economic* effects of such controls. Further, although researchers such as Cousins (see Cousins (1992)) and Scoones (see Scoones (1992, 1993, 1995, 1996) and Scoones and Cousins (1994)) have provided valuable accounts of a number of ecological and economic aspects of range management in Southern Africa, their accounts are (i) largely descriptive and not theoretical in nature and (ii) not explicitly about the theoretical properties of spatial and temporal controls. As such, we reiterate that our point is not that there are no ecological-economic studies of range management but that — as the title of this chapter suggests — there are virtually no theoretical ecological-economic studies that analyze the formal or mathematical properties of spatial and temporal controls and the ways in which these controls impact the ecology and the economics of rangelands.

Given this state of affairs, the objective of this chapter is to identify and delineate a research agenda concerning four outstanding questions in range management. Answers to these questions will not only fill a gap in the extant theoretical literature but they will also shed valuable light on practical issues concerning range management in general.

However, before we move to the questions themselves let us first say a few things about why these questions collectively are more interesting and more relevant than the general question "What is the best rangeland grazing management regime?" We recognize that whether a particular question is more or less interesting than another question is, at least to some extent, in the eyes of the beholder. Having said this, the reader should note that we are not saying that any one of the following questions, viewed in isolation, is more interesting than the above general question. What we are saying is that the four questions of this chapter, viewed *collectively*, are not only more interesting from a research perspective but also more relevant for practical range management in Southern Africa and elsewhere in the world. Why do we say this? There are two reasons. First, experience with range management in Southern Africa, North America, and Australia tells us clearly that there is no single "best rangeland grazing management regime." As Smith and Pickup (1993) have noted, although there are some general principles that can be applied across continents, effective range management is context dependent. Consequently, if range managers are to be successful then they must pay attention to the particularities of local situations. Put differently, because we do not believe that there is one "best rangeland grazing management regime," it follows that we do not believe it is particularly helpful to focus on the above general question.

---

7 Examples of such studies include Huffaker *et al.* (1989), Torrell *et al.* (1991), and McCluskey and Rausser (1999).

8 Examples include Graetz (1986), Gutman *et al.* (1999), and Rehman *et al.* (1999).

Second, although the principal management issue on communal grazing lands in Southern Africa involves the prevention of rangeland degradation[9] — and we have already noted that the prevention of rangeland degradation is a salient concern of short duration grazing — range managers need to be cognizant of the fact that there are differences between countries — management practices that work in Botswana will not necessarily work in South Africa — and differences within individual countries — stocking policies that are appropriate for Eastern Botswana are not always appropriate for rangelands in other parts of Botswana — in Southern Africa. From a management standpoint, this means that we need to study *flexible* grazing systems with attributes that can be applied to many geographical areas within, say, Southern Africa. Short duration grazing is a flexible grazing system with features — such as rotation — that are common to and hence relevant for many parts of Southern Africa. The four questions that follow discuss some of these common and relevant features. As such, we believe that the collective knowledge gained from researching the four questions — the research agenda — of this chapter will provide us with more useful knowledge about flexible management practices than will a focus on the general question "What is the best rangeland grazing management regime?" We now discuss the four individual questions.

2. The Optimal Number of Paddocks

The first question that we wish to address relates to the spatial control aspect of range management.[10] Specifically, the question of interest is this: What is the optimal number of paddocks for rangelands that are managed using short duration (or cell) grazing? There are numerous applied analyses of this question. For instance, Walker and Heitschmidt (1986) have compared two cell grazing systems, the first with 14 cells and the second with 42 cells. However, one can ask why we should compare 14 versus 42 cells and not, say, 17 versus 39 cells? Walker and Heitschmidt (1986) provide no theoretical basis for their choice of 14 and 42 cells. Similarly, Holechek *et al.* (2001, p. 269) recommend "that no fewer than eight pastures (paddocks) of equal grazing capacity be built that radiate as spokes from the central area where water is located." Once again, one can ask: What is the basis for recommending that the number of paddocks be eight? Holechek *et al.* (2001) provide no answer.

Indeed, it is fair to say that these cited studies are representative of the extant literature on this question. Consequently, theoretical research is needed to answer this question. One novel way to proceed would be to use queuing theory[11] to provide a stochastic characterization of a rangeland with paddocks that is managed using short duration grazing. Once this task has been accomplished, it will be possible to show that the optimal number of paddocks is the solution to an optimization problem — such as an expected net social benefit maximization problem — faced by a range manager.

---

9 See Abel (1993) and Behnke and Scoones (1993) for additional details on this point.

10 For more on veld management in South Africa with paddocks (also called camps) of different numbers and sizes, see Wilcox (1976). For a general discussion of the salience of spatial considerations in range management in Southern Africa, see Smith and Pickup (1993).

11 For more on queuing theory, see Kulkarni (1995) and Ross (1996, 2000).

3. The Optimal Temporal Control

With reference to Zimbabwe, Savory and Butterfield (1999, p. 41, emphasis in original) have noted that until "recently no one truly explored the question of *when* animals are [on the range] as opposed to how many there are." They then go on to argue that overgrazing bears "little relationship to the number of animals but rather to the *time* plants [are] exposed to the animals" (Savory and Butterfield, 1999, p. 46, emphasis in original).

This leads to the second question on which theoretical research is needed and this question concerns the temporal control aspect of rangeland management. In particular, how should a range manager set the temporal control optimally for rangelands that are managed using short duration grazing? As with the first question discussed in the previous section, it is easy to find an abundance of applied but not theoretical studies of this question. For the most part, these studies have focused on the effects of temporal controls in specific regions. For instance, in an interesting early study in South Africa,[12] Acocks (1966) analyzed the effects of concentrating animals on a small portion of the rangeland, allowing these animals to graze all the plants evenly, and then moving them to another portion of the rangeland. Weltz and Wood (1986) have analyzed the water infiltration impacts of short duration grazing in central New Mexico. Similarly, Pluhar *et al.* (1987) have studied the effect that short duration grazing has on water infiltration and on sediment production in the Texas plains. Finally, Manley *et al.* (1997) have examined the forage production and vegetation composition effects of short duration grazing in Wyoming.

Because these and other similar studies provide no theoretical basis for the determination of the temporal control in short duration grazing, we know *precious little* about the *properties* of optimal temporal controls. Consequently, it would be very useful to use a dynamic and stochastic framework, such as a renewal-theoretic framework,[13] to delineate a range manager's objective function. This objective function might be the long run expected net cost function. Once this has been done, it will be possible to show — either in general or with the assistance of specific functional forms — that the optimal length of the grazing period in a paddock (the temporal control) can be chosen to minimize, for instance, the above mentioned long run expected net cost function.

4. The Steady State Distribution of Animals

With regard to Zimbabwe and Southern Africa, Scoones (1996) and Smith and Pickup (1993) have pointed out that effective range management involves decision making under *uncertainty*. The same point has also been made by Holechek *et al.* (2001). For instance, there is uncertainty about climate and there is uncertainty about the effects of diseases on animals. Despite the salience of uncertainty in determining the efficacy of short duration grazing, there are very few studies of short duration grazing in a *stochastic* framework.[14] In a seminal paper,

---

[12] The term "short duration grazing" was not used in 1966. However, the Acocks (1966) paper greatly influenced Allan Savory and his writings on short duration grazing. For more on this, see Savory and Butterfield (1999, pp. 44-46).

[13] Lucid accounts of renewal theory can be found in Ross (1996, 2000).

[14] There are numerous applied analyses of short duration grazing in a *deterministic* context. For more on these studies, the reader should consult Pluhar *et al.* (1987), Skovlin (1987), and Manley *et al.* (1997).

Westoby *et al.* (1989) described but did not model the effects of uncertainty on the temporal evolution of rangelands. Similarly, Ellis *et al.* (1993) have commented on range management problems stemming from climatic variability in the Turkana district of Northern Kenya. More recently, Batabyal *et al.* (2001, also chapter 8) have compared the optimal stocking rate with the optimal length of time during which animals graze a specific parcel of rangeland. Finally, Batabyal and Godfrey (2002, also chapter 7) have provided a conceptual approach to rangeland management under uncertainty.

These papers have certainly advanced our understanding of some aspects of range management in both deterministic and stochastic settings. Nevertheless, we still know very little about the stochastic properties of short duration grazing. Specifically, we do *not* know the answer to the following important question: What is the steady state distribution of animals for rangelands that are managed using short duration grazing? The answer to this question is important because of at least two reasons. First, for rangelands managed using short duration grazing, this answer can be used to determine the impact of alternate kinds of uncertainty on managerial decisions about optimal spatial and temporal controls. Second, once one knows the steady state distribution of animals, one can begin to assess the distribution of rangeland degradation. In the context of Botswana in particular and Southern Africa in general, Abel (1993) has observed that it is important to perform this assessment of the distribution of rangeland degradation.

Given this state of affairs, an appropriate stochastic framework is needed to pose and answer the above question. In this context, the theory of alternating renewal processes[15] can be used to effectively analyze the question of interest. This suggested framework has two desirable features. First, using this framework, it will *not* be necessary to use specific functional forms for the cumulative distribution functions to model the random amount of time that the animals spend in the various paddocks of a rangeland. Second, it will be possible to model the amount of time spent by the animals in distinct paddocks with *dissimilar* cumulative distribution functions.

5. The Effect of Potential Irreversibilities

Rangelands can exist in a number of different states. In the simplest case, rangelands that have not been overgrazed can exist in a "desirable" state in which there is substantial forage cover. On the other hand, it is now well known that overgrazed rangelands can exist in an "undesirable" desertified state.[16] Generalizing from this simple example, a rangeland can reasonably be thought to exist in not just two but a finite number of states. Some of these states are desirable and others are undesirable. As well, the reader should note that in both these sets of states, some states are better than others. Now, limiting attention to the undesirable set of states, what is important for our purpose is that some states are likely to be *irreversible*. In these irreversible or *crisis* states,[17] the rangeland is so degraded — forage cover on the rangeland is virtually non-existent — that no matter how hard a manager might try, (s)he will be unable to move the rangeland to any other state. Given this state of affairs,

---

15 For readable accounts of alternating renewal processes, see Kulkarni (1995) and Ross (1996, 2000).

16 For more on these issues see Jodha (1980), Ho (2000), Batabyal and Godfrey (2002, also chapter 7), and Batabyal (2003).

17 In the rest of this chapter, we shall use the terms "irreversible state" and "crisis state" interchangeably.

one way to look at the task of range management is to say that a range manager's aim is to *maximize* the amount of time the rangeland spends in the desirable set of states. Using the same logic, one can also say that a range manager's task is to *minimize* the amount of time the rangeland spends in the undesirable — but not the irreversible — set of states.

Now, all range managers operate in a *probabilistic* environment. In other words, the state of a managed rangeland at any particular point in time is a function not only of the actions undertaken by this manager but also of uncertain environmental factors such as droughts, fires, and pest infestations. What this means is that even though the range manager believes that (s)he is taking actions to ensure that the rangeland does not hit any one of the crisis states, the rangeland *may* still do so. Put differently, despite the range manager's best efforts, (s)he can never be sure that the rangeland under study will not hit an irreversible state.

Given this state of affairs, how should a manager go about the task of managing a rangeland? In particular, what are the properties of the following reasonable approach: First, identify the crisis state or states. Although it is generally not possible to do so exactly, on the basis of past experience and available scientific information, range managers are often able to specify a safe minimum standard,[18] and all the states of the rangeland below the safe minimum standard can be thought of as crisis states. Next, it is necessary to put in place a well designed plan of action. In the context of a specific rangeland, this might mean the use of a specialized grazing system such as short duration grazing.[19]

As discussed in the previous two paragraphs, even with a well designed plan of action in place, it is still *possible* that a managed rangeland will hit a crisis state. Accordingly, a key question now relates to the (random) length of time until the crisis state is reached. Put differently, the range manager would like to know how long it takes for the rangeland to hit a crisis state. In addition, what does the answer to this query depend on? Finally, is the answer history dependent or independent? In other words, does the answer to the length of time until crisis query depend on the state in which the range manager's plan of action is put in place? As best as we can tell, there are no previous analyses of these important questions.

Consequently, the first step in analyzing the above questions is to delineate a dynamic and stochastic rangeland mathematically. This can be done effectively by modeling a dynamic and stochastic rangeland with a discrete-time Markov chain.[20] Some of this chain's states will be transient and other states — the crisis state(s) — will be absorbing. Once this has been done, the outstanding task will be to use a technique such as first step analysis (see Taylor and Karlin (1998)) to determine the expected amount of time it takes to hit the crisis state(s) for alternate range management regimes.

This research is needed to determine, *inter alia*, whether there are practical rules of thumb that range managers can use to ascertain when a managed rangeland is close to an irreversible or crisis state. Such knowledge will help individuals manage rangelands whose behavior over time is typically governed by a great deal of unpredictability.

---

[18] For more on the practical aspects of range management under uncertainty, see Campbell and Stafford Smith (2000). For a general discussion of the use of the safe minimum standard in the presence of potential irreversibilities, see Farmer and Randall (1998).

[19] For more on short duration grazing, see Batabyal (2001, 2002, also chapter 9), Holechek *et al.* (2001), and Batabyal and Yoo (2002, also chapter 5).

[20] Textbook accounts of discrete-time Markov chains can be found in Ross (1996, 2000) and in Taylor and Karlin (1998).

6. CONCLUSIONS

The theoretical study of range management from an ecological-economic perspective calls for the adoption of a unified approach. In particular, to address this chapter's four salient issues, this unified approach will need to pay equal attention to the ecological and to the economic aspects of the problem of comprehending the dynamic and the stochastic behavior of rangelands. If this unified approach is adopted, then we can expect to see two main results.

First, the research community will have a much better idea about the behavior of jointly determined rangelands whose evolution is governed by ecological and economic factors. Second, in the context of the integrated approach to range management that we have described, social welfare is an increasing function of the health of rangelands. In turn, the health of a rangeland depends, *inter alia*, on the policies that a range manager has in place to ensure that the goods and the services provided by rangelands are utilized judiciously. By focusing on the optimal management of rangelands, the research that we have suggested be undertaken will shed light on how changes in rangeland health translate into changes in social welfare.

REFERENCES

Abel, N.O.J. 1993. Reducing Cattle Numbers on Southern African Communal Range: Is it Worth it? in R.H. Behnke, I. Scoones, and C. Kerven, (Eds.), *Range Ecology at Disequilibrium*. Overseas Development Institute, London, UK.

Acocks, J.P. 1966. Non-Selective Grazing as a Means of Veld Reclamation, *Proceedings of the Grassland Society of Southern Africa*, 1, 33-39.

Batabyal, A.A. 2001. Some Theoretical Aspects of Short Duration Grazing, *International Journal of Ecology and Environmental Sciences*, 27, 215-220.

Batabyal, A.A. 2002. An Upper Bound for the Transient Resilience of a Rangeland that is Managed Using Cell Grazing, *Journal of Economic Research*, 7, 151-160.

Batabyal, A.A. 2003. A Note on First Step Analysis and Rangeland Management Under Uncertainty, *Unpublished Manuscript*, Rochester Institute of Technology.

Batabyal, A.A., Biswas, B., and Godfrey, E.B. 2001. On the Choice Between the Stocking Rate and Time in Range Management, *Environmental and Resource Economics*, 20, 211-223.

Batabyal, A.A., and Godfrey, E.B. 2002. Rangeland Management Under Uncertainty: A Conceptual Approach, *Journal of Range Management*, 55, 12-15.

Batabyal, A.A., and Yoo, S.J. 2002. The Steady State Distribution of Animals in Short Duration Grazing, *Applied Economics Letters*, 9, 695-698.

Behnke, R.H., and Scoones, I. 1993. Rethinking Range Ecology: Implications for Rangeland Management in Africa, in R.H. Behnke, I. Scoones, and C. Kerven, (Eds.), *Range Ecology at Disequilibrium*. Overseas Development Institute, London, UK.

Behnke, R.H., Scoones, I, and Kerven, C. Eds. 1993. *Range Ecology at Disequilibrium*. Overseas Development Institute, London, UK.

Campbell, B.D., and Stafford Smith, D.M. 2000. A synthesis of Recent Global Change Research on Pasture and Rangeland Production: Reduced Uncertainties and Their Management Implications, *Agriculture, Ecosystems, and Environment*, 82, 39-55.

Cousins, B. 1992. *Managing Communal Rangeland in Zimbabwe*. Commonwealth Secretariat, London, UK.

Ellis, J.E., Coughenour, M.B., and Swift, D.M. 1993. Climate Variability, Ecosystem Stability, and the Implications for Range and Livestock Development, in R.H. Behnke, I. Scoones, and C. Kerven, (Eds.), *Range Ecology at Disequilibrium*. Overseas Development Institute, London, UK.

Farmer, M.C., and Randall, A. 1998. The Rationality of a Safe Minimum Standard, *Land Economics*, 74, 287-302.

Graetz, R.D. 1986. A Comparative Study of Sheep Grazing in a Semi-Arid Saltbush Pasture in Two Condition Classes, *Australian Rangeland Journal*, 8, 46-56.

Gutman, M., Holzer, Z., Baram, H., Noy-Meir, I., and Seligman, N.G. 1999. Heavy Stocking and Early-Season Deferment of Grazing on Mediterranean-Type Grassland, *Journal of Range Management*, 52, 590-599.

Hartwick, J.M., and Olewiler, N.D. 1998. *The Economics of Natural Resource Use*, 2rd edition. Addison-Wesley, Reading, MA.

Ho, P. 2000. China's Rangelands Under Stress: A Comparative Study of the Pasture Commons in the Ninxia Hui Autonomous Region, *Development and Change*, 31, 385-412.

Holechek, J.L., Pieper, R.D., and Herbel, C.H. 2001. *Range Management*, 4th edition. Prentice Hall, Upper Saddle River, NJ.

Huffaker, R.G., Wilen, J.E., and Gardner, B.D. 1989. Multiple Use Benefits on Public Rangelands: An Incentive-Based Fee System, *American Journal of Agricultural Economics*, 71, 670-678.

Jodha, N.S. 1980. The Process of Desertification and the Choice of Interventions, *Economic and Political Weekly*, 15, 1351-1356.

Karpoff, J.M. 1987. Suboptimal Controls in Common Resource Management: The Case of the Fishery, *Journal of Political Economy*, 95, 179-194.

Kulkarni, V.G. 1995. *Modeling and Analysis of Stochastic Systems*. Chapman and Hall, London, UK.

Manley, W.A., Hart, R.H., Samuel, M.J., Smith, M.A., Waggoner, Jr., J.W., and Manley, J.T. 1997. Vegetation, Cattle, and Economic Responses to Grazing Strategies and Pressure, *Journal of Range Management*, 50, 638-646.

Matulich, S.C., Mittelhammer, R.C., and Roberte, C. 1996. Toward a More Complete Model of Individual Transferable Fishing Quotas: Implications of Incorporating the Processing Sector, *Journal of Environmental Economics and Management*, 31, 112-128.

McCluskey, J.J., and Rausser, G.C. 1999. Federal Grazing Reform and Avoidable Risk, *Journal of Agricultural and Resource Economics*, 24, 140-154.

Perrings, C., and Walker, B. 1995. Biodiversity Loss and the Economics of Discontinuous Change in Semiarid Rangelands, in C. Perrings, K.G. Maler, C. Folke, C.S. Holling, and B.O. Jansson, (Eds.), *Biodiversity Loss*. Cambridge University Press, Cambridge, UK.

Pluhar, J.J., Knight, R.W., and Heitschmidt, R.K. 1987. Infiltration Rates and Sediment Production as Influenced by Grazing Systems in the Texas Rolling Plains, *Journal of Range Management*, 40, 240-244.

Rehman, A. Mackintosh, J.B., Warren, B.E., and Lindsay, D.R. 1999. Revegetated Saline Pastures as a Forage Reserve for Sheep: 1. Effects of Season and Grazing on Morphology and Nutritive Value of Saltbush, *Rangeland Journal*, 21, 3-12.

Ross, S.M. 1996. *Stochastic Processes*, 2nd edition. Wiley, New York, NY.

Ross, S.M. 2000. *Introduction to Probability Models*, 7th edition. Academic Press, San Diego, CA.

Savory, A. 1983. The Savory Grazing Method or Holistic Resource Management, *Rangelands*, 5, 155-159.

Savory, A., and Butterfield, J. 1999. *Holistic Management*, 2nd edition. Island Press, Washington, DC.

Scoones, I. 1992. Coping With Drought: Responses of Herders and Livestock in Contrasting Savanna Environments in Southern Zimbabwe, *Human Ecology*, 20, 293-314.

_____ . 1993. Why are there so Many Animals? Cattle Population Dynamics in the Communal Areas of Zimbabwe, in R.H. Behnke, I. Scoones, and C. Kerven, (Eds.), *Range Ecology at Disequilibrium*. Overseas Development Institute, London, UK.

_____ . 1995. Exploiting Heterogeneity: Habitat Use by Cattle in the Communal Areas of Zimbabwe, *Journal of Arid Environments*, 29, 221-237.

_____ . 1996. *Hazards and Opportunities*. Zed Books, London, UK.

Scoones, I., and Cousins, B. 1994. The Struggle for Control over Wetland Resources in Zimbabwe, *Society and Natural Resources*, 7, 579-594.

Skovlin, J.M. 1987. Southern Africa's Experience With Intensive Short-Duration Grazing, *Rangelands*, 9, 162-168.

Smith, M.S., and Pickup, G. 1993. Out of Africa, Looking In: Understanding Vegetation Change, in R.H. Behnke, I. Scoones, and C. Kerven, (Eds.), *Range Ecology at Disequilibrium*. Overseas Development Institute, London, UK.

Taylor, H.M., and Karlin, S. 1998. *An Introduction to Stochastic Modeling*, 3rd edition. Academic Press, San Diego, CA

Torell, L.A., Lyon, K.S., and Godfrey, E.B. 1991. Long Run versus Short Run Planning Horizons and the Rangeland Stocking Rate Decision, *American Journal of Agricultural Economics*, 73, 795-807.

Walker, J.W., and Heitschmidt, R.K. 1986. Effect of Various Grazing Systems on Type and Density of Cattle Trails, *Journal of Range Management*, 39, 428-431.

Weltz, M., and Wood, M.K. 1986. Short-Duration Grazing in Central New Mexico: Effects on Infiltration Rates, *Journal of Range Management*, 39, 365-368.

Weninger, Q., and Strand, I.E. 1998. Regulating Production Time to Control Output: An Application to a Harvest Technology, *Unpublished Manuscript*, Utah State University.

Westoby, M., Walker, B., and Noy-Meir, I. 1989. Opportunistic Management for Rangelands Not at Equilibrium, *Journal of Range Management*, 42, 266-274.

Wilcox, D.G. 1976. *Aspects of Veld Management in South Africa and South West Africa*. Rangeland Management Branch, Department of Agriculture, Canberra, Australia.

Young, O.R. 1981. *Natural Resources and the State*. University of California Press, Berkeley, CA.

Chapter 3

THE OPTIMAL NUMBER OF PADDOCKS IN SHORT DURATION GRAZING

ABSTRACT

This chapter addresses the following previously unstudied question in range management: What is the optimal number of paddocks for rangelands that are managed using short duration grazing? We first use queuing theory to provide a novel stochastic characterization of a rangeland with paddocks that is managed using short duration grazing. Next, we use this characterization to show that the optimal number of paddocks is the solution to an expected profit maximization problem faced by a range manager.

Keywords: Dynamics, Paddock, Profit Maximization, Range Management, Uncertainty

1. INTRODUCTION

Fisheries across North America and in many other parts of the world are commonly managed with season length restrictions (a temporal control) and with area restrictions (a spatial control). Although this fact is now well known (see Hartwick and Olewiler (1998)), what is less well known is that this two-pronged regulatory approach is also used to manage rangelands. Holechek *et al.* (2001) have pointed out that deferred rotation grazing, the Merrill three herd and four pasture system, rest rotation grazing, and particularly short duration grazing (also called cell grazing), all have spatial and temporal control aspects to them. However, the temporal and particularly the *spatial* control features of range management can be seen most clearly in the case of short duration grazing.

In this form of grazing, rangelands are managed by controlling the number of paddocks into which a given parcel of rangeland is divided (a spatial control) and by controlling the length of time during which animals graze the individual paddocks (a temporal control).[1] As noted in Holechek *et al.* (2001), the basic idea in short duration grazing is threefold. First, the

[1] Readers interested in learning more about short duration grazing should consult Goodloe (1969), Savory and Parsons (1980), Savory (1983), Weltz and Wood (1986), Pluhar *et al.* (1987), Savory and Butterfield (1999), Holechek *et al.* (2001), Batabyal *et al.* (2001a, also chapter 8, 2001b), and Batabyal and Yoo (2002, also chapter 5).

manager divides the rangeland under consideration into a number of fenced paddocks, either with or without a wagon wheel design. Next, the manager allows his herd of animals to graze a particular paddock for a specific time period. Finally, the herd of animals is moved to the next paddock, and the manager continues this process in sequential fashion. For rangelands managed with short duration grazing, it has been said that "a well-managed cell permits stocking rates to be increased substantially...compared to...other grazing systems" (Holechek *et al.*, 2001, p. 271).

There are numerous *applied* analyses of short duration grazing. For instance, Skovlin (1987) has studied the effects of short duration grazing on Southern African grasslands. Pluhar *et al.* (1987) have examined the effects that short duration grazing has on water infiltration and on sediment production in the Texas plains. Finally, Manley *et al.* (1997) have analyzed the forage production and vegetation composition effects of short duration grazing in Wyoming.

These papers have certainly increased our understanding of the region specific impacts of short duration grazing. Nevertheless, because there are virtually no *theoretical* studies of this kind of grazing, we know very little about the procedures for selecting the optimal number of paddocks in short duration grazing. For instance, Walker and Heitschmidt (1986) have compared two cell grazing systems, the first with 14 cells and the second with 42 cells. However, one can ask why we should compare 14 versus 42 cells and not, say, 17 versus 39 cells? Walker and Heitschmidt provide no theoretical basis for their choice of 14 and 42 cells. Similarly, Holechek *et al.* (2001, p. 269) recommend "that no fewer than eight pastures (paddocks) of equal grazing capacity be built that radiate as spokes from the central area where water is located." Once again, one can ask: What is the basis for recommending that the number of paddocks be eight? No answer is provided by these authors.

Given this state of affairs, our point is this: Even though spatial and temporal controls have been and are used in short duration grazing, we know very *little* about how a manager should go about the task of selecting the *optimal* number of paddocks in a rangeland managed with short duration grazing. As such, the objective of this chapter is to shed light on the following hitherto *unstudied* question about short duration grazing: What is the optimal number of paddocks for rangelands that are managed using short duration grazing? We use queuing theory — to the best of our knowledge for the first time — to provide a novel stochastic characterization of a rangeland with paddocks that is managed using short duration grazing.[2] Next, we use this characterization to show that the optimal number of paddocks is the solution to an expected profit maximization problem faced by a manager.

The rest of this chapter is organized as follows: Section 2 uses a queuing theoretic framework to delineate a rangeland with paddocks that is managed using short duration grazing. Section 3 poses our manager's expected profit maximization problem and then discusses this problem and its solution. Section 4 concludes and offers suggestions for future research in range management.

---

2 As indicated previously, there are three basic steps in short duration grazing. First, the manager divides the rangeland under consideration into a number of fenced paddocks, either with or without a wagon wheel design. Next, the manager allows his herd of animals to graze a particular paddock for a specific time period. Finally, the herd of animals is moved to the next paddock, and the manager continues this process in sequential fashion. This three step process can be modeled well with a particular type of closed network queue, the so called cyclic queue (see Gross and Harris (1985), Kulkarni (1995), and Ross (2000)). This is why we're using queuing theory to study the optimal number of paddocks in short duration grazing.

2. A Stochastic Short Duration Grazing System

2.1. Preliminaries

Consider a parcel of rangeland that has been divided into fenced paddocks, either with or without a wagon wheel design. We shall index these paddocks by i where $i = 1,...,N$. The manager has a finite number of animals and in what follows, we shall refer to this finite number of animals as the herd of animals.

The manager begins grazing his herd of animals in a particular paddock, say the ith paddock, and the animals graze this paddock for an exponentially distributed amount of time with rate $\mu_i(1)$, where the one in parentheses refers to the fact that we're studying one herd of animals. After the completion of grazing in paddock i the manager moves his herd of animals to paddock $i+1$. The animals then graze paddock $i+1$ for an exponential amount of time with rate $\mu_{i+1}(1)$.[3] Upon completion of grazing in paddock $i+1$ the manager moves his animals to paddock $i+2$ and he continues in this fashion until the original state (state i) is reached. This sequential movement between the different paddocks of the rangeland can be compactly described by using the notion of a transition probability.[4] Specifically, for our N paddock grazing system, we have

$$p_{i,i+1} = 1, \quad 1 \le i \le N-1, \text{ and } p_{N,1} = 1. \tag{1}$$

Equation (1) tells us that upon completion of grazing in paddock i the herd of animals will next graze paddock $i+1$ with certainty. Also, when the last paddock N is reached, the manager will complete the grazing cycle by moving his herd of animals to paddock 1 with certainty. The reader should note three things. First, these transition probabilities can also be expressed as the elements of the transition probability matrix $P = [p_{ij}]$ Second, the Markov chain (also called the embedded Markov chain;[5] see Kulkarni (1995) and Ross (2000) for additional details) associated with this matrix P will typically have unique steady state (also called stationary) probabilities. Third, we can express these steady state probabilities by the vector $\delta = (\delta_1, \delta_2, ..., \delta_N)$ where δ_i, $i = 1,...,N$ is the steady state probability that the Markov chain is in state i and δ solves

$$\delta_j = \sum_{i=1}^{i=N} \delta_i p_{ij}, \text{ and } \sum_{j=1}^{j=n} \delta_j = 1. \tag{2}$$

---

[3] Strictly speaking, if the manager's herd consists of α animals, then we should express the rates as $\alpha\mu_i(1)$, $\alpha\mu_{i-1}(1)$ etc., and not as stated in the text. However, this does not alter the analysis substantively and hence, in what follows, we shall not be concerned with this point.

[4] For more on transition probabilities, see Kulkarni (1995) and Ross (2000).

[5] Exponential queuing models are special examples of continuous-time Markov chains (CTMCs) and hence they can be analyzed using the methods used to study CTMCs. By looking at a CTMC at certain points in time and by choosing these time points so as to exploit the memoryless property of the exponential distribution, it is possible to identify the embedded Markov chain.

The events described in the previous two paragraphs can be modeled well with a particular type of closed network queue, the so called cyclic queue.[6] In a cyclic queue, neither customers (the animals) nor servers (the paddocks) are allowed to enter or leave the queuing system (our N paddock grazing system). Further, the customers (the animals) move from one server (paddock) to the next in a cyclical fashion so that the output of one server (say paddock i) is the input to the next server (paddock $i+1$) and in particular, the output of the last server (paddock N) is the input to the first server (paddock 1). Now, to provide an answer to the central question of this chapter, let us compute the steady state or stationary probabilities of this cyclic queue (N paddock grazing system).

2.2. Computation of the Steady State Probabilities

Let $X_i(t)$ denote the number of animals in paddock i at time t and let $X(t)=\{X_1(t),X_2(t)...,X_N(t)\}$ denote the state of the N paddock grazing system. Note that in our case, the number of animals in any paddock *i, i=1,...N,* is constant and equals the size of the manager's herd. This means that the state space of our N paddock grazing system is given by

$$S=\{x=(x_1,x_2,...x_i,...x_N):\sum_{i=1}^{i=N} x_i=1\} \tag{3}$$

where x_i is a N-vector with 0's everywhere and a 1 in the ith position. Now, it is well known that the stochastic process $\{X(t):t\geq 0\}$ is a continuous-time Markov chain for which a steady state probability distribution function exists (see Kulkarni (1995)). Consequently, our task now is to compute the relevant steady state or stationary probabilities.

In the context of our simple N paddock grazing system, equation (2) yields

$$\delta_1=\delta_2=...=\delta_N=1/N. \tag{4}$$

Equation (4) tells us that the steady state probabilities of the embedded Markov chain of our N paddock grazing system are given by the reciprocal of the total number of paddocks into which the rangeland under study has been divided. Now using equation 7.131 in Kulkarni (1995) or equation 8.22 in Ross (2000) we conclude that the steady state probabilities of the N paddock grazing system that we are after are

$$p(x_i)=\frac{1/\mu_i(1)}{\sum_{j=1}^{j=N} 1/\mu_j(1)} \text{ and } \sum_{i=1}^{i=N} p(x_i)=1 \tag{5}$$

---

[6] Cyclic queues are an important subset of closed network queues. To learn more about cyclic queues, see Gross and Harris (1985), Kulkarni (1995), and Ross (2000).

Because the length of the grazing period in any paddock $i, i = 1,...,N,$ is an exponentially distributed random variable with rate $\mu_i(1),$ we know that the mean length of the grazing period in paddock i is $1/\mu_i(1)$. Using this information, it is straightforward to interpret the first part of equation (5). This part tells us that the steady state probability of the manager's herd of animals being in paddock $i, i = 1,...,N,$ equals the mean length of the grazing period in paddock i divided by the sum of the mean lengths of the grazing periods in all the N paddocks that the rangeland under study has been divided into. The second part of equation (5) simply says that when summed over all the paddocks, these steady state probabilities equal one. What is the optimal value of N? We are now in a position to answer this question.

3. Choosing the Number of Paddocks Optimally

3.1. The Expected Profit Maximization Problem

We suppose that our manager's objective is to maximize profits from his range operations.[7] To this end, short duration grazing on the manager's rangeland gives rise to revenues and to costs. These revenues and costs depend on the number of paddocks (N) and on the mean lengths of the grazing period $(1/\mu_i(1))$ in the individual paddocks. Let us denote the revenue and the cost from grazing in paddock i by the functions $R_i\{N,1/\mu_i(1)\}$ and $C_i\{N,1/\mu_i(1)\}$ respectively. We assume that $R_i\{\cdot,\cdot\}$ is strictly increasing and strictly concave in its arguments for all i and that $C_i\{\cdot,\cdot\}$ is strictly increasing and strictly convex in its arguments for all i. The manager's problem is to choose N to maximize his expected profit. In other words, this manager solves

$$\max_N \sum_{i=1}^{i=N}[p(x_i)[R_i\{N,1/\mu_i(1)\} - C_i\{N,1/\mu_i(1)\}]], \tag{6}$$

where the $p(x_i)$'s are given by equation (5). Let us now examine four salient features of this expected profit maximization problem.

3.2. Discussion

First, note that along with N if we make the $\mu_i(1)$'s choice variables, then this problem can be used to simultaneously determine the optimal number of paddocks and the optimal mean length of the grazing period in each paddock. Viewed in this way, problem (6) is a mixed integer programming problem because the $\mu_i(1)$'s are continuous control variables and N, the optimal number of paddocks, is an integer control variable.

---

7 It is also possible to analyze a scenario in which the manager's objective is to minimize the costs of range operations. However, because profit maximization implies cost minimization and cost minimization does not imply profit maximization, the profit maximization focus of this chapter is more appropriate.

Second, if we are willing to approximate the optimal integer N with the optimal continuous N then we can use calculus to determine the optimal continuous N. In this case, the optimal number of paddocks is N^* where

$$N^* = \arg\max_N [\sum_{i=1}^{i=N} [p(x_i)[R_i\{N,1/\mu_i(1)\} - C_i\{N,1/\mu_i(1)\}]]]$$

However, if we do adopt this approximation approach, then it is important to recognize that the optimal continuous N will not always be close to the optimal integer N (see Bradley *et al.* (1977), Schrijver (1998), and Wolsey (1998)).

Third, if the above continuous approximation is either infeasible or undesirable, then formal integer programming techniques (see Bradley *et al.* (1977), Schrijver (1998), and Wolsey (1998)) will have to be used to solve the manager's problem (6).

Finally, the reader will note that problem (6) is an unconstrained expected profit maximization problem. In a given range management setting, it is quite possible that the specific maximization problem facing the manager will involve one or more constraints. If this is the case then these constraints will have to be incorporated into the problem along the lines suggested in Bradley *et al.* (1977), Schrijver (1998), and Wolsey (1998), and the optimal N will be the solution to this constrained maximization problem.

4. Conclusions

This chapter contains the *first* application of queuing theory to an important problem in rangeland management. We addressed the following specific question: What is the optimal number of paddocks for rangelands that are managed using short duration grazing? After noting that this question has not received prior theoretical attention in the range management literature, we used a queuing theoretic framework to provide a novel stochastic characterization of a rangeland with paddocks that is managed using short duration grazing. Next, we used this characterization to show that the optimal number of paddocks is the solution to an expected profit maximization problem faced by a manager.

The analysis contained in this chapter can be extended in a number of directions. In what follows, we suggest two possible extensions of this chapter's research. First, on the theoretical front, recall that in this chapter the manager's animals graze a paddock for an exponentially distributed amount of time. The research of this chapter can be extended by analyzing a queuing model in which the length of the grazing period in each paddock is not exponentially but arbitrarily distributed. Second, on the empirical front, it would be helpful to collect data and to see if the estimated revenue and cost functions satisfy the concavity and the convexity assumptions that we've made in this chapter. Analyses of rangeland management that incorporate these features of the problem into the analysis will provide further insights into the functioning of rangelands that are managed using short duration grazing.

REFERENCES

Batabyal, A.A., Biswas, B., and Godfrey, E.B. 2001a. On the Choice Between the Stocking Rate and Time in Range Management, *Environmental and Resource Economics*, 20, 211-223.

_____ . 2001b. The Stocking Rate Versus Time in Range Management: A Review of Recent Theoretical Developments, *Applied Economics Letters*, 8, 555-559.

Batabyal, A.A., and Yoo, S.J. 2002. The Steady State Distribution of Animals in Short Duration Grazing, *Applied Economics Letters*, 9, 695-698.

Bradley, S.P., Hax, A.C., and Magnanti, T.L. 1977. *Applied Mathematical Programming*. Addison-Wesley, Reading, MA.

Goodloe, S. 1969. Short Duration Grazing in Rhodesia, *Journal of Range Management*, 22, 369-373.

Gross, D., and Harris, C.M. 1985. *Fundamentals of Queuing Theory*, 2nd edition. Wiley, New York, NY.

Hartwick, J.M., and Olewiler, N.D. 1998. *The Economics of Natural Resource Use*, 2nd edition. Addison-Wesley, Reading, MA.

Holechek, J.L., Pieper, R.D., and Herbel, C.H. 2001. *Range Management*, 4th edition. Prentice-Hall, Upper Saddle River, NJ.

Kulkarni, V.G. 1995. *Modeling and Analysis of Stochastic Systems*. Chapman and Hall, London, UK.

Manley, W.A., Hart, R.H., Samuel, M.J., Smith, M.A., Waggoner, Jr., J.W., and Manley, J.T. 1997. Vegetation, Cattle, and Economic Responses to Grazing Strategies and Pressure, *Journal of Range Management*, 50, 638-646.

Pluhar, J.J., Knight, R.W., and Heitschmidt, R.K. 1987. Infiltration Rates and Sediment Production as Influenced by Grazing Systems in the Texas Rolling Plains, *Journal of Range Management*, 40, 240-244.

Ross, S.M. 2000. *Introduction to Probability Models*, 7th edition. Harcourt Academic Press, San Diego, CA.

Savory, A. 1983. The Savory Grazing Method or Holistic Resource Management, *Rangelands*, 5, 155-159.

Savory, A., and Butterfield, J. 1999. *Holistic Management*. Island Press, Washington, DC.

Savory, A., and Parsons, S.D. 1980. The Savory Grazing Method, *Rangelands*, 2, 234-237.

Schrijver, A. 1998. *Theory of Linear and Integer Programming*. Wiley Interscience, New York, NY.

Skovlin, J.M. 1987. Southern Africa's Experience With Intensive Short-Duration Grazing, *Rangelands*, 9, 162-168.

Walker, J.W., and Heitschmidt, R.K. 1986. Effect of Various Grazing Systems on Type and Density of Cattle Trails, *Journal of Range Management*, 39, 428-431.

Weltz, M., and Wood, M.K. 1986. Short Duration Grazing in Central New Mexico: Effects on Infiltration Rates, *Journal of Range Management*, 39, 365-368.

Wolsey, L.A. 1998. *Integer Programming*. Wiley, New York, NY.

Chapter 4

AN APPROACH TO THE CHOICE OF THE OPTIMAL TEMPORAL CONTROL IN SHORT DURATION GRAZING

ABSTRACT

The purpose of this chapter is to discuss an approach to the choice of the optimal temporal control in range management. To this end, suppose that an individual (the manager) is concerned about the ecological and the economic aspects of range management. Further, suppose that this manager has decided to use short duration grazing to manage the rangeland under consideration. In this scenario, we address the following specific question: How should this range manager choose the temporal control optimally? We use a renewal-theoretic framework to first derive our manager's long run mean net cost function. We then demonstrate that the optimal temporal control minimizes this criterion function.

Keywords: Cost Function, Grazing, Optimal Temporal Control, Renewal Theory, Uncertainty

1. INTRODUCTION

Several researchers such as Young (1981), Hartwick and Olewiler (1998), and Batabyal and Beladi (2002) have pointed out that it is not at all uncommon for renewable resources such as fisheries to be managed with spatial and temporal controls. For instance, fisheries in the United States Pacific Northwest are frequently managed with season length stipulations (one kind of temporal control) and with geographic regulations (one kind of spatial control).

Batabyal *et al.* (2001a, also chapter 8) have noted that the combination of spatial and temporal controls has been used to manage rangelands as well. In this context, Holechek *et al.* (2001, pp. 254-277), have remarked that specialized grazing systems such as deferred rotation grazing, the Merrill three-herd and four-pasture system, rest rotation grazing, and particularly short duration grazing,[1] all have spatial and temporal control features to them. This notwithstanding, it is fair to say that the spatial and the temporal control features of range management can be seen most clearly in the case of short duration grazing.

In this form of grazing, rangelands are managed by controlling both the length of time during which animals graze a paddock (this is the temporal control) and by controlling the area of this paddock (this is the spatial control). The essential idea in short duration grazing is threefold. First, the manager divides the rangeland or cell under consideration into a number of fenced paddocks, either with or without a wagon wheel design. Second, the manager permits his or her herd of animals to graze a particular paddock for a specific time period. Third, the herd of animals is transferred to the next paddock, and the manager continues this process in sequential fashion. For rangelands managed with short duration grazing, it has been said that "a well-managed cell permits stocking rates to be increased substantially...compared to...other grazing systems" (Holechek *et al.*, 2001, p. 271).

In addition to ascertaining the optimal length of the time period that the herd of animals should spend on a particular paddock, a question of some significance confronting range managers concerns the determination of the optimal stocking rate. This query has already been analyzed in Batabyal *et al.* (2001a, also chapter 8, 2001b) and therefore we shall not reflect on this issue in this chapter.

It is now well known that there are myriad *applied* analyses of short duration grazing. Here are three examples: Weltz and Wood (1986) have examined the water infiltration impacts of short duration grazing in central New Mexico. In like manner, Pluhar *et al.* (1987) have investigated the impacts that short duration grazing has on water infiltration and on sediment production in the Texas plains. Finally, Manley *et al.* (1997) have explored the forage production and vegetation composition effects of short duration grazing in Wyoming. Without a doubt, these papers have increased our comprehension of the location-specific impacts of short duration grazing. Even so, because there are no theoretical analyses of short duration grazing, we don't know a whole lot about the formal characteristics of short duration grazing. In this context, our specific point is this: Despite the fact that temporal controls have been and are utilized in short duration grazing, we know very little about the theoretical properties of *optimal* temporal controls in short duration grazing.

Given this situation, the purpose of this chapter is to shed light on the following query regarding short duration grazing that has received scant attention in the literature: How should a range manager choose the temporal control optimally for a rangeland that is managed using short duration grazing? To answer this question, we use a renewal-theoretic framework[2] to derive the range manager's long run mean net cost function. We then demonstrate that the optimal temporal control minimizes this criterion function. The remainder of this chapter is organized as follows: Section 2 uses a renewal-theoretic model to derive our range manager's long run mean net cost function. Next, this section discusses several features of the solution to our manager's long run cost minimization problem. Finally, with the aid of two examples, this section shows how the temporal control can be chosen to minimize the range manager's long run mean net cost function in specific circumstances. Section 3 concludes and offers recommendations for future research in range management.

---

[1] This specialized system is also known as cell grazing, rapid rotation grazing, and time controlled grazing.

[2] Clear introductions to renewal theory can be found in Kulkarni (1995), Ross (1996, 2000), and Taylor and Karlin (1998).

2. CHOOSING THE TEMPORAL CONTROL OPTIMALLY

2.1. Preliminaries

Consistent with standard practice in short duration grazing (see Holechek *et al.*, 2001, pp. 269-270), we assume that the dynamic and stochastic rangeland (the cell) under study has been divided into a number of fenced paddocks of equal grazing capacity.[3] Given the stocking rate versus time controversy in the range management literature (see Savory and Butterfield (1999), Holechek *et al.* (2000), and Batabyal *et al.* (2001a, also chapter 8, 2001b, 2001c, also chapter 12)), our goal in this chapter is to focus on the choice of the optimal *temporal* control. This is consistent with the remark that overgrazing bears "little relationship to the number of animals but rather to the *time* plants [are] exposed to the animals" (Savory and Butterfield, 1999, p. 46, emphasis in original).

In short duration grazing, the rangeland or cell under study is divided into a number of paddocks and these paddocks represent the spatial control variable for our range manager. In other words, during the pertinent grazing season, one prominent task performed by the range manager involves sequentially moving his or her herd of animals from one paddock to the next. Because these fenced paddocks are generally of equal grazing capacity (see Holechek *et al.* (2001, p. 269) and Sevi *et al.* (2001)), the fundamental question now concerns the determination of the manager's temporal control. In other words, how long should his or her herd of animals graze a particular paddock before the herd is moved to the next paddock? As Batabyal (2001b) has noted, this query has great significance for practical range management. Indeed, the success of short duration grazing is contingent on choosing this temporal control variable optimally. In spite of this, there are very few theoretical studies of this question. In the words of Holechek *et al.* (2001, p. 270), this lack of research has precluded one from "drawing very many definite conclusions about the effectiveness of various short-duration grazing strategies." Given this situation, we now use a renewal-theoretic framework to derive our range manager's long run mean net cost function. We then show that the optimal temporal control minimizes this criterion function.

2.2. The Renewal-Theoretical Framework

The renewal-theoretic framework of this paper is adapted from Batabyal (2001b) and Batabyal and Beladi (2002). Without loss of generality, consider an arbitrary paddock of the rangeland under study. Is forage quality on a paddock a function of the grazing pressure applied to this paddock? A number of papers have addressed this question. In particular, Holechek *et al.* (1999) have demonstrated that forage production across North America typically averaged 23% higher under moderate than heavy grazing and 36% higher under light than heavy grazing. In addition to this, the work of Cook and Harris (1950), Parker (1969), DeBano and Rice (1973), and Thurow *et al.* (1986) has shown us that the levels of phosphorus, protein, cell solubles, and the presence of water affect the growth rates of forage

[3] One question that arises at this point relates to the *number* of paddocks. In other words, into how many paddocks should the rangeland under study be divided? This question has recently been analyzed by Batabyal (2001a, also chapter 3). Consequently, we omit a discussion of this issue.

plants in different ways. For instance, in the Great Plains of the United States, forage grows and is nourishing for anywhere between 140 and 180 days. In contrast, in the western United States, forage is actively growing and salubrious for only 60 to 90 days in a year. From the perspective of this chapter, the consequence of all this is that "[f]orage...quality on most ranges varies tremendously between seasons" (Holechek *et al.*, 2001, p. 318). We model this feature of the problem by supposing that forage quality in the paddock under study is a *random* variable.

Now, in order to be compatible with the units of measurement, we assume that there exists a one-to-one and strictly monotone function that maps the forage quality of a paddock to what we shall call its operative lifetime. The forage quality of an ungrazed paddock is dependent on environmental factors only. Put differently, such a paddock has the highest operative lifetime. As this paddock is grazed for increasingly longer periods of time, its forage quality and hence its operative lifetime diminishes stochastically.

Note that forage quality is a random variable. Moreover, operative lifetime is a one-to-one and strictly monotonic transformation of forage quality. Therefore, operative lifetime is itself a random variable. Let us denote the distribution and the density functions of operative lifetime by $G(\bullet)$ and $g(\bullet)$ respectively. In the rest of this chapter, we shall not refer to the forage quality of a paddock. Instead, we shall talk about its operative lifetime. Even so, the reader should understand that any particular value of operative lifetime refers clearly to a unique level of forage quality. Further, like all lifetimes, operative lifetime is measured in time units and it is defined over the non-negative real numbers.

The range manager will stop grazing in a particular paddock and transfer his or her herd of animals to the next paddock either when T time periods have passed, or when a random and deleterious environmental event curtails the operative lifetime of this paddock, thereby leading to a breakdown of this paddock as far as subsequent grazing is concerned. Examples of such injurious environmental events include unusually high rainfall in an arid paddock that results in significant topsoil loss and the appearance of blood sucking insects that stress cattle on the paddock.[4]

The culmination of grazing in a distinct paddock and the simultaneous movement of the herd of animals to the next paddock results in ecological and economic benefits and costs to our manager. The ecological and the economic benefits stem from things like forage recovery and animal weight gain, respectively.[5] In turn, the ecological and the economic costs arise from things like the detrimental impacts of the hoof actions of the manager's animals on water infiltration on the paddock and from feeding and fencing, respectively.[6] Let us denote the ecological and the economic net cost to our manager from the culmination of grazing on a paddock after T time periods, i.e., after one grazing cycle by $c_T(ecol, econ)$. If grazing on a paddock has to be terminated because of the occurrence of a injurious environmental event, i.e., if a breakdown occurs, then we suppose that our manager incurs a different kind of ecological and economic net cost given by $c_B(ecol, econ)$. This net cost accounts for the fact that when a damaging environmental event occurs, expenditure will generally be incurred by our manager on things such as the application of fertilizers to the paddock soil, the repair of

---

[4] This is a frequent problem in high mountain meadow ranges in the United States. See Holechek *et al.* (2001, p. 421) for additional details.

[5] See Holechek and Stephenson (1983) and Holechek *et al.* (2001).

[6] See Warren *et al.* (1986), Batabyal (2001b), and the references cited therein.

existing fences, and the potential replacement of dead or diseased animals. Our assignment now is to determine the range manager's long run mean net cost function.

As indicated previously, to compute this cost function, we shall use a renewal-theoretic framework in general and the so called renewal-reward theorem in particular.[7] To apply this theorem to our problem, we will first need to determine the expected net cost per grazing cycle. To this end, let L denote the operative lifetime of a paddock and let C denote the net cost per grazing cycle. Then, given the description of net costs in the previous paragraph, it is clear that the net cost per grazing cycle is

$$C = \begin{cases} c_T(ecol, econ) \text{ if } L > T \\ c_B(ecol, econ) \text{ if } L \leq T. \end{cases} \quad (1)$$

The expectation of this net cost, $E[C]$, can be obtained by noting that

$$E[C] = c_T(ecol, econ)Pr\{L > T\} + c_B(ecol, econ)Pr\{L \leq T\}. \quad (2)$$

The distribution function of the random variable L is $G(\bullet)$. From this we can deduce that

$$E[C] = c_T(ecol, econ)\{1 - G(T)\} + c_B(ecol, econ)G(T). \quad (3)$$

The next step in our derivation of the manager's long run mean net cost function involves computing the expected amount of time it takes to complete a grazing cycle. To calculate this expectation, observe that the length of a grazing cycle on the paddock is T if $L > T$ and it is L if $L \leq T$. This tells us that the expectation we are interested in is $E[\min(L,T)]$. Now because both L and T are non-negative random variables, we can compute the expected amount of time it takes to complete a grazing cycle by using the apposite tail distribution. From Wolff (1989, p. 36) it follows that

$$E[\min(L,T)] = \int_0^T \{1 - G(l)\}dl, \quad (4)$$

where l is a realization of the operative lifetime random variable L.

We are now in a position to state our manager's long run mean net cost function. By the renewal-reward theorem (see Wolf (1989, p. 60)), this criterion function is given by dividing equation (3) by equation (4). Performing this division, we get

$$Long\ Run\ Mean\ Net\ Cost = \frac{c_T(ecol, econ)\{1 - G(T)\} + c_B(ecol, econ)G(T)}{\int_0^T \{1 - G(l)\}dl}. \quad (5)$$

---

[7] For more on this theorem, see Wolff (1989, p. 60) and Ross (2000, p. 378).

2.3. Discussion

Equation (5) delineates our range manager's criterion function. In particular, this manager's objective is to choose *T,* the length of a grazing cycle on the paddock, to minimize the long run mean net cost of management operations when these operations have ecological and economic consequences. The solution to the above minimization problem can be written as

$$T^* = argmin_T \left[\frac{c_T(ecol,econ)\{1 - G(T)\} + c_B(ecol,econ)G(T)}{\int_0^T \{1 - G(l)\}dl} \right]. \tag{6}$$

Examining equation (6), we see that the optimal temporal control T^* depends on $c_T(ecol,econ)$ and on $c_B(ecol,econ)$. This clearly demonstrates the functional dependence of the optimal length of the grazing cycle on the ecological and the economic features of the management problem.

We now use two examples to show how the theoretical approach of this chapter might be operationalized. In the first example, the operative lifetime random variable L is distributed uniformly. In the second example, L is distributed exponentially.

2.3.1. An Example with Uniformly Distributed L

In hundreds of dollars, suppose $c_T(ecol,econ) = \$1, c_B(ecol,econ) = \$5,$ and $L \sim U(0,1)$. Then, using this information, the fact that $G(l) = l$ for $0 \le l \le 1,$ and equation (5), we find that our manager's long run mean net cost equals $(2 + 8T)/(2T - T^2)$. Consequently, this manger's task now is to solve

$$\min_{\{T\}} \left[\frac{2 + 8T}{2T - T^2} \right]. \tag{7}$$

The first order necessary condition to this problem is given by the quadratic equation

$$2T^2 + T - 1 = 0. \tag{8}$$

The reader can confirm that the second order sufficiency condition is satisfied. The solutions to equation (8) are $T = 1/2$ and $T = -1$. Because $T \ge 0,$ it is clear that the optimal value of T is $T^* = 1/2$. Consequently, in this first example, if we measure time in days, then our manager's optimal policy is as follows: Graze the paddock for half a day, then discontinue grazing on this paddock and move the herd of animals to the next paddock, and carry on in this manner. We now proceed to our second example.

2.3.2. An Example with Exponentially Distributed L

As in the previous section, once again we assume that $c_T(ecol, econ) = \$1$ and $c_B(ecol, econ) = \$5$, both costs being measured in hundreds of dollars. However, in this example, the operative lifetime random variable L is exponentially distributed. This means that $G(l) = 1 - e^{-\lambda l}$, $\lambda > 0$. Now, using this information, we can compute our manager's long run mean net cost. After some algebra, we see that this cost equals $5\lambda + \lambda/(e^{\lambda T} - 1)$. As such, our manger's task now is to solve

$$\min_{\{T\}} [5\lambda + \frac{\lambda}{e^{\lambda T} - 1}]. \tag{9}$$

Inspecting equation (9), we discern two salient features of this minimization problem. First, in the limit, our manager's long run mean net cost approaches infinity as T approaches zero. Second, also in the limit, this long run mean net cost approaches 5λ, its lowest possible value, as T approaches infinity. Hence it is clear that in this exponential case, the optimal value of T is $T^* = \infty$. This means that the optimal length of the grazing cycle is infinitely long. The reader should note that this unusual result arises entirely because of our assumption that the operative lifetime random variable L is exponentially distributed. Indeed, with exponential lifetimes, no matter how long a paddock has been grazed, it is as if this paddock had never been grazed.[8]

A range manager who chooses the optimal temporal control T^* in the manner described in the previous two examples will achieve two goals simultaneously. First, he or she will guarantee that range operations are conducted at the *lowest* mean net cost to society. Second, because our manager's focus is on the *long run* and because this manager pays attention to the ecological and to the economic features of the management problem (see equations (5) and (6)), by choosing the temporal control in the manner just specified, our manager will also make sure that the rangeland under study is used sustainably. In other words, when used in this way, the rangeland will provide society with a flow of consumptive and non-consumptive services on a sustained basis.

3. Conclusions

We analyzed an issue in this chapter that has received scant attention in the extant literature on short duration grazing. In particular, we addressed the following query: How should a range manager — who employs a long run viewpoint and who pays attention to the ecological and to the economic features of the management problem — set the temporal control optimally for a rangeland that is managed using short duration grazing? To answer this question, we used a renewal-theoretic framework and the renewal-reward theorem in particular to derive the range manager's long run mean net cost function. We then discussed several features of the solution to this manager's long run cost minimization problem. Finally,

---

8 For more on this memoryless feature of the exponential distribution, see Taylor and Karlin (1998, pp. 35-36).

with the aid of two examples, we showed how the temporal control can be chosen optimally to minimize the above mentioned criterion function in specific circumstances.

The analysis contained in this chapter can be extended in a number of directions. We now suggest two possible extensions of this chapter's research. First, this chapterer's research can be extended by analyzing a situation in which the operative lifetimes of distinct paddocks are given by dissimilar cumulative distribution functions. To this end, on the empirical front, it would be helpful to collect data on the costs and to determine which cumulative distribution functions are most useful in modeling the "operative lifetime" random variable. Studies of range management that incorporate these features of the problem into the analysis will provide further insights into the functioning of rangelands that are managed using short duration grazing.

REFERENCES

Batabyal, A.A. 2001a. On the Optimal Number of Paddocks in Short Duration Grazing, *Journal of Economic Research*, 6, 159-169.

_____ . 2001b. Some Theoretical Aspects of Short Duration Grazing, *International Journal of Ecology and Environmental Sciences*, 27, 215-220.

Batabyal, A.A., and Beladi, H. 2002. On the Choice of the Optimal Temporal Control in Renewable Resource Management, *Stochastic Environmental Research and Risk Assessment*, 16, 325-332.

Batabyal, A.A., Biswas, B., and Godfrey, E.B. 2001a. On the Choice Between the Stocking Rate and Time in Range Management, *Environmental and Resource Economics*, 20, 211-223.

_____ . 2001b. The Stocking Rate Versus Time in Range Management: A Review of Recent Theoretical Developments, *Applied Economics Letters*, 8, 555-559.

_____ . 2001c. A Dynamic and Stochastic Perspective on the Role of Time in Range Management, *Unpublished Manuscript*, Rochester Institute of Technology.

Cook, C.W., and Harris, L.E. 1950. The Nutritive Content of the Grazing Sheep's Diet on the Summer and Winter Ranges of Utah, *Utah Agricultural Experiment Station Bulletin 342*.

DeBano, L.F., and Rice, R.M. 1973. Water-Repellent Soils: Their Implications in Forestry, *Journal of Forestry*, 71, 220-223.

Hartwick, J.M., and Olewiler, N.D. 1998. *The Economics of Natural Resource Use*, 2nd edition. Addison Wesley, Reading, MA.

Holechek, J.L., and Stephenson, T. 1983. Comparison of Big Sagebrush Vegetation in Northcentral New Mexico Under Moderately Grazed and Grazing Excluded Conditions, *Journal of Range Management*, 36, 455-457.

Holechek, J.L., Gomes, H., Molinar, F., and Galt, D. 1999. Grazing Studies: What We've Learned, *Rangelands*, 21, 12-16.

Holechek, J.L., Gomes, H., Molinar, F., Galt, D., and Valdez, R. 2000. Short-Duration Grazing: The Facts in 1999, *Rangelands*, 22, 18-23.

Holechek, J.L., Pieper, R.D., and Herbel, C.H. 2001. *Range Management*, 4th edition. Prentice-Hall, Upper Saddle River, NJ.

Kulkarni, V.G. 1995. *Modeling and Analysis of Stochastic Systems*. Chapman and Hall, London, UK.

Manley, W.A., Hart, R.H., Smith, M.A., Waggoner, J.W., and Manley, J.T. 1997. Vegetation, Cattle, and Economic Responses to Grazing Strategies and Pressure, *Journal of Range Management*, 50, 638-646.

Parker, K.G. 1969. The Nature and Use of Utah Range, *Utah State University Extension Circular 359*, Logan, UT.

Pluhar, J.J., Knight, R.W., and Heitschmidt, R.K. 1987. Infiltration Rates and Sediment Production as Influenced by Grazing Systems in the Texas Rolling Plains, *Journal of Range Management*, 40, 240-244.

Ross, S.M. 1996. *Stochastic Processes*, 2nd edition. Wiley, New York, NY.

_____ . 2000. *Introduction to Probability Models*, 7th edition. Harcourt Academic Press, San Diego, CA.

Savory, A., and Butterfield, J. 1999. *Holistic Management*, 2nd edition. Island Press, Washington, DC.

Sevi, A., Muscio, A., Dantone, D., Iascone, V., and D'Emilio, F. 2001. Paddock Shape Effects on Grazing Behavior and Efficiency in Sheep, *Journal of Range Management*, 54, 122-125.

Taylor, H.M., and Karlin, S. 1998. *An Introduction to Stochastic Modeling*, 3rd edition. Academic Press, San Diego, CA.

Thurow, T.L., Blackburn, W.H., and Taylor, C.A. 1986. Hydrologic Characteristics of Vegetation Types as Affected by Livestock Grazing Systems, Edwards Plateau, Texas, *Journal of Range Management*, 39, 505-508.

Warren, S.D., Blackburn, W.H., and Taylor, C.A. 1986. The Influence of Livestock Trampling Under Intensive Rotation Grazing on Soil Hydrologic Conditions, *Journal of Range Management*, 39, 491-496.

Weltz, M., and Wood, M.K. 1986. Short-Duration Grazing in Central New Mexico: Effects on Infiltration Rates, *Journal of Range Management*, 39, 365-368.

Wolff, R.W. 1989. *Stochastic Modeling and the Theory of Queues*. Prentice-Hall, Englewood Cliffs, NJ.

Young, O.R. 1981. *Natural Resources and the State*. University of California Press, Berkeley, CA.

Chapter 5

The Steady State Distribution of Animals in Short Duration Grazing

with S. J. Yoo

Abstract

This chapter addresses the following hitherto unstudied question in rangeland management: What is the steady state distribution of livestock animals for rangelands that are managed using short duration grazing? We use the theory of alternating renewal processes to answer this question and our answer has two desirable features. First, we do not assume a specific functional form for the cumulative distribution function of the amount of time that the animals spend in the various paddocks. Second, we permit the amount of time spent by the animals in distinct paddocks to be represented by dissimilar cumulative distribution functions.

Keywords: Alternating Renewal Process, Paddock, Short Duration Grazing, Uncertainty

1. Introduction

Specialized grazing systems have been used to manage rangelands in North America at least since the 1950s. The management of rangelands with these grazing systems typically involves the use of spatial and temporal controls. Spatial controls tell us which pastures can be grazed at a particular point in time and temporal controls specify the length of time during which animals may graze a particular pasture. Indeed, as Holechek *et al.* (2001) have pointed out, deferred rotation grazing, the Merrill three herd and four pasture system, rest rotation grazing, and particularly short duration grazing (also called cell grazing), all have spatial and temporal control aspects to them. However, the spatial and the temporal control features of range management can be seen most clearly in the case of short duration grazing.

In this form of grazing, rangelands are managed by controlling the number of paddocks into which a given parcel of rangeland is divided (a spatial control), and by controlling the length of time during which animals graze the individual paddocks (a temporal control).[1] The central idea in short duration grazing is threefold. First, the manager divides the rangeland under consideration into a number of fenced paddocks, either with or without a wagon-wheel design. Second, the manager allows his herd of animals to graze a particular paddock for a specific time period. Third, the herd of animals is moved to the next paddock, and the manager continues this process in sequential fashion. For rangelands managed with short duration grazing, it has been said that "a well-managed cell permits stocking rates to be increased substantially...compared to...other grazing systems" (Holechek *et al.*, 2001, p. 271).

Holechek *et al.* (2001) and others have noted that effective range management involves decision making under *uncertainty*. There is uncertainty about climate, there is uncertainty about the effects of diseases on livestock animals, and particularly in the case of short duration grazing, there is uncertainty about the outcomes of specific decisions such as the determination of the length of time during which animals graze the paddocks of the rangeland under study. Despite the salience of uncertainty in determining the efficacy of short duration grazing, there are very few studies of short duration grazing in a *stochastic* framework.[2]

In a fundamental paper, Westoby *et al.* (1989) described but did not model the effects of uncertainty on the temporal evolution of rangelands. More recently, Batabyal *et al.* (2001, also chapter 8) have analyzed range management in a stochastic framework. However, the emphasis of this paper is not on short duration grazing *per se* but on comparing the optimal stocking rate with the optimal length of time during which animals graze a specific parcel of rangeland. Finally, Batabyal and Godfrey (2002, also chapter 7) have provided a conceptual approach to rangeland management under uncertainty. However, the focus of this last paper is on range management in general and not on short duration grazing in particular.

These papers have certainly advanced our understanding of range management in both deterministic and stochastic settings. Nevertheless, because there are virtually no studies of short duration grazing in the presence of uncertainty, we know very little about the stochastic properties of this kind of grazing. Specifically, we do *not* know the answer to the following important question: What is the steady state distribution of livestock animals for rangelands that are managed using short duration grazing? The answer to this question is important because — for rangelands managed using short duration grazing — it can be used to determine the optimal number of paddocks that a given parcel of rangeland should be divided into and to determine the optimal length of the grazing period in a particular paddock.

Given this state of affairs, the purpose of this chapter is to use the theory of alternating renewal processes[3] to answer the question posed in the previous paragraph. The reader will see that our answer has two desirable features. First, we do *not* assume a specific functional form for the cumulative distribution function of the amount of time that the animals spend in the various paddocks of the rangeland. Second, we permit the amount of time spent by the animals in distinct paddocks to be represented by *dissimilar* cumulative distribution functions.

The rest of this chapter is organized as follows: Section 2 first uses a renewal theoretic framework to delineate a rangeland with paddocks that is managed using short duration

[1] Readers interested in learning more about short duration grazing should consult Savory and Butterfield (1999), Batabyal *et al.* (2001, also chapter 8), and Holechek *et al.* (2001).

[2] There are numerous applied analyses of short duration grazing in a *deterministic* context. For more on these studies, the reader should see Pluhar *et al.* (1987), Skovlin (1987), and Manley *et al.* (1997).

[3] For readable accounts of alternating renewal processes, see Kulkarni (1995) and Ross (1996, 2000).

grazing. Next, this section computes the steady state distribution of livestock animals for this rangeland. Finally, this section discusses the connections between this steady state distribution and the optimal number of paddocks, and the optimal length of the grazing period in a paddock. Section 3 concludes and offers suggestions for future research in range management.

2. A STOCHASTIC CELL GRAZING SYSTEM

2.1. Preliminaries

Consider a parcel of rangeland that has been divided into C fenced paddocks, either with or without a wagon-wheel design. We shall index these paddocks by i, where $i = 1,...,C$. The range manager has a finite number of animals and in what follows, we shall refer to this finite number of animals as the herd of animals.

The range manager begins grazing his herd of animals in paddock 1, and the animals graze this first paddock for a stochastic length of time with a general cumulative distribution function $F_1(\cdot)$. After the completion of grazing in paddock 1, the manager moves his herd of animals to paddock 2. The animals then graze paddock 2 for a stochastic length of time with a general cumulative distribution function $F_2(\cdot)$ Upon completion of grazing in paddock 2, the manager moves his herd of animals to paddock 3 and he continues in this fashion until he reachers paddock C. From paddock C, the animals return to paddock 1 and then the grazing process starts over.[4] What is important for our purpose is that the herd of animals graze paddock $i, i = 1,...,C$, for a stochastic length of time with a general cumulative distribution function $F_i(\cdot)$. The reader will note that we have allowed the amount of time spent by the animals in distinct paddocks to be represented by *dissimilar* cumulative distribution functions. Our task now is to compute, for any paddock $i, i = 1,...,C$, the steady state probability that the manager's herd of animals is grazing that paddock. Mathematically, we wish to compute $\lim_{t\to\infty} Prob\{herd\ is\ grazing\ paddock\ i\ at\ time\ t\}$

2.2. Computing the Steady State Probabilities

The amount of time spent by the manager's herd of animals in any paddock $i, i = 1,...,C$, is a non-negative random variable. Consequently, following Ross (1996), we can represent the mean of this random variable in paddock i by

$$\theta_i = \int_0^\infty \overline{F}_i(t)dt \text{ where } \overline{F}_i(\cdot) = F_i(\cdot). \quad (1)$$

[4] For technical reasons, we suppose that $G(\cdot)$, the cumulative distribution function of the time between entrances to paddock 1, is non-lattice and that it has a finite mean. For more on lattice random variables, see Kulkarni (1995) and Ross (1996, 2000).

In words, equation (1) tells us that the mean of a non-negative random variable can be expressed as the integral from 0 to of this variable's tail distribution function.

Let us say that the short duration grazing process under study is "on" when the herd of animals is grazing any of the paddocks *1,2,...,i,* and that it is "off" otherwise. Now, using equation (1) and either theorem 8.23 in Kulkarni (1995) or theorem 3.4.4 in Ross (1996), we have

$$\lim_{t\to\infty} Prob\{herd\ is\ grazing\ any\ of\ paddocks\ 1,...,i\} = \frac{\theta_1 + \theta_2 + ... + \theta_i}{\sum_{j=1}^{j=C} \theta_j} \quad (2)$$

Equation (2) gives us the steady state probability that the manager's herd of animals is grazing any of the paddocks 1 through *i.* However, we are interested in determining the steady state probability that the herd of animals is grazing paddock *i.* This required probability can be obtained by subtraction. Specifically, we have

$$\begin{aligned} &\lim_{t\to\infty} Prob\{herd\ is\ grazing\ paddock\ i\ at\ time\ t\} = \\ &\lim_{t\to\infty} Prob\{herd\ is\ grazing\ any\ of\ paddocks\ 1,...,i\} - \\ &\lim_{t\to\infty} Prob\{herd\ is\ grazing\ paddock\ \leq i-1\}. \end{aligned}$$

Now, using equations (1), (2), and the preceding equality, we get

$$\lim_{t\to\infty} Prob\{herd\ is\ grazing\ paddock\ i\ at\ time\ t\} = \frac{\theta_i}{\sum_{j=1}^{j=C} \theta_j}, i = 1,...,C \quad (3)$$

Despite the apparent complexity of alternating renewal processes, equation (3) tells us that the steady state distribution of livestock animals in short duration grazing is given by a simple ratio. For an arbitrary paddock, say paddock *i,* equation (3) tells us that the steady state probability that the manager's herd of animals is grazing paddock i equals the mean amount of time spent grazing this paddock divided by the sum of the mean times spent grazing all the C paddocks that the rangeland has been divided into.

2.3. Discussion

Equation (3) contains the answer to the question posed in section 1 of this chapter. As indicated previously, this answer is important for at least two reasons. First, observe that the question of the optimal number of paddocks that a given parcel of rangeland should be divided into essentially boils down to choosing the integer C optimally. Because C is an argument in the right hand side (RHS) of equation (3), this equation can be used to construct an appropriate objective function for our manager. One way to determine the optimal number of paddocks would be to choose C to optimize this objective function subject to one or more constraints.

Second, from equation (3) we see that the steady state distribution of livestock animals is a function of the mean times (the θ_i's) spent grazing the individual paddocks. Consequently, this fact can also be used to formulate a stochastic control problem for our manager. If the manager chooses the θ_i's to optimize the objective function of this control problem then he will have set the length of the grazing period in each paddock optimally.

3. Conclusions

This chapter contains the *first* application of the theory of alternating renewal processes to a salient problem in rangeland management. We addressed the following specific question: What is the steady state distribution of livestock animals for rangelands that are managed using short duration grazing? After noting that this question has not received prior theoretical attention in the range management literature, we used a renewal theoretic framework to provide a novel stochastic representation of a rangeland with paddocks that is managed using short duration grazing. Next, we used this representation to compute the steady state distribution of livestock animals for rangelands that are managed using short duration grazing.

As indicated in section 2.3, the analysis contained in this chapter can be extended in a number of directions. In particular, it would be useful to use equation (3) to formulate a stochastic control problem for a range manager that can be used to jointly determine the optimal number of paddocks and the optimal length of the grazing period in a paddock. A resolution of these issues will undoubtedly increase our understanding of the behavior of rangelands that are managed using short duration grazing.

References

Batabyal, A.A., B. Biswas, and E.B. Godfrey. 2001. On the Choice Between the Stocking Rate and Time in Range Management, *Environmental and Resource Economics*, 20, 211-223.

Batabyal, A.A., and Godfrey, E.B. 2002. Rangeland Management Under Uncertainty: A Conceptual Approach, *Journal of Range Management*, 55, 12-15.

Holechek, J.L., Pieper, R.D., and Herbel, C.H. 2001. *Range Management*, 4th ed. Prentice-Hall, Upper Saddle River, NJ.

Kulkarni, V.G. 1995. *Modeling and Analysis of Stochastic Systems*. Chapman and Hall, London, UK.

Manley, W.A., Hart, R.H., Samuel, M.J., Smith, M.A., Waggoner, Jr., J.W., and Manley, J.T. 1997. Vegetation, Cattle, and Economic Responses to Grazing Strategies and Pressure, *Journal of Range Management*, 50, 638-646.

Pluhar, J.J., Knight, R.W., and Heitschmidt, R.K. 1987. Infiltration Rates and Sediment Production as Influenced by Grazing Systems in the Texas Rolling Plains, *Journal of Range Management*, 40, 240-244.

Ross, S.M. 1996. *Stochastic Processes*, 2nd ed. John Wiley and Sons, New York, NY.

_____ . 2000. *Introduction to Probability Models*, 7th ed. Harcourt Academic Press, San Diego, CA.

Savory, A., and Butterfield, J. 1999. *Holistic Management*, 2nd ed. Island Press, Washington DC.

Skovlin, J.M. 1987. Southern Africa's Experience With Intensive Short-Duration Grazing, *Rangelands*, 9, 162-168.

Westoby, M., Walker, B., and Noy-Meir, I. 1989. Opportunistic Management for Rangelands Not at Equilibrium, *Journal of Range Management*, 42, 266-274.

Chapter 6

POTENTIAL IRREVERSIBILITIES IN RANGE MANAGEMENT OVER TIME AND UNDER UNCERTAINTY

ABSTRACT

A fundamental objective of rangeland managers all over the world is to take measures to ensure that the rangeland being managed stays away from irreversible states in which it provides little or no consumptive and non-consumptive services to humans. Nevertheless, despite a manager's best attempts, the rangeland under study may still hit an irreversible state. Accordingly, given a particular management regime, it is useful to know the length of time taken to hit an irreversible state. In this chapter, we conduct a theoretical analysis of this hitherto unstudied question. We first provide a probabilistic description of two range management regimes. Next, we calculate the mean time until an irreversible state is hit for both these regimes. Finally, we provide a numerical example to demonstrate the working of our model and then we discuss the ramifications of our findings for rangeland management in general.

Keywords: Dynamics, Irreversibility, Management, Rangeland, Uncertainty

1. INTRODUCTION

Extensive present-day discussions about sustainable development have brought the subject of renewable resource management into sharp focus (Dasgupta and Maler (1997)). Although there is some disagreement among scholars about the exact meaning of the term "sustainable development," there is general agreement on the point that if the process of development is to be sustainable, then essential renewable resources such as rangelands will need to be optimally managed.

It has rightly been observed that irrespective of "whose figures are used, rangelands are the primary land type in the world" (Holechek *et al.*, 2001, p. 9). In addition, because of two reasons, rangelands are an important renewable resource. First, they perform a number of vital ecological functions such as nutrient cycling, production of organic matter, and the decomposition of organic matter. Second, rangelands provide humans with economic goods and services including food, fiber, building materials, and wildlife viewing. As a result, range managers frequently attempt to manipulate "range components to obtain the optimum combination of goods and services for society on a sustained basis" (Holechek *et al.*, 2001, p. 5).

In many developing nations, rangelands are a vital renewable resource. This is because people in developing countries are largely agrarian and pastoral in nature. As such, they often rely on rangelands for their well-being. For instance, villagers living in the watershed of the Alaknanda river in the central Himalayas in India spent 25% of their total working time collecting fuel, caring for animals, and grazing their animals (Center for Science and Environment (1990)). This dependence on rangelands is not limited to South Asia. Consider the case of Africa. At least 15 to 20 million pastoralists live in the arid and the semiarid regions of Africa. These pastoralists "derive most of their income and sustenance from livestock grazing..." (Holechek *et al.*, 2001, p. 533). Given the nature and the extent of this dependence on rangelands in both developing and developed countries,[1] it should be clear to the reader that the optimal management of this resource is a very important policy objective. Indeed, in this context, it is useful to point out that environmental problems are "almost always associated with resources that are regenerative but are in danger of exhaustion from excessive use" (Dasgupta, 1996, p. 389).

Rangelands can exist in a number of different states. In the simplest case, rangelands that have not been overgrazed can exist in a "desirable" state in which there is substantial forage cover. On the other hand, it is now well known that overgrazed rangelands can exist in an "undesirable" desertified state.[2] Generalizing from this simple example, a rangeland can reasonably be thought to exist in not just two but a finite number of states. Some of these states are desirable and others are undesirable. As well, the reader should note that in both these sets of states, some states are better than others. Now, limiting attention to the undesirable set of states, what is significant for our purpose is that some states are likely to be *irreversible*. In these irreversible or *crisis* states,[3] the rangeland is so degraded — forage cover on the rangeland is virtually non-existent — that no matter how hard a manager might try, (s)he will be unable to move the rangeland to any other state. Given this state of affairs, one way to look at the task of rangeland management is to say that a range manager's aim is to *maximize* the amount of time the rangeland spends in the desirable set of states. Using the same logic, one can also say that a range manager's task is to *minimize* the amount of time the rangeland spends in the undesirable — but not the irreversible — set of states.

Now, all range managers operate in a *probabilistic* environment. In other words, the state of a managed rangeland at any particular point in time is a function not only of the actions undertaken by this manager but also of imprecise environmental factors such as droughts, fires, and pest infestations. What this means is that even though the range manager believes that (s)he is taking actions to ensure that the rangeland does not hit any one of the crisis states, the rangeland *may* still do so. Put differently, despite the range manager's best efforts, (s)he can never be sure that the rangeland under study will not hit an irreversible state.

---

1 For a more detailed corroboration of this claim see Heady and Heady (1982), Sandford (1983), and Holechek *et al.* (2001).

2 For more on these issues see Jodha (1980), Ho (2000), Batabyal and Godfrey (2002, also chapter 7), and Batabyal (2003).

3 In the rest of this chapter, we shall use the terms "irreversible state" and "crisis state" interchangeably.

Given this state of affairs, how should a manager go about the task of managing a rangeland? We now analyze the attributes of the following reasonable approach. First, identify the crisis state or states. Although it is generally not possible to do so exactly, on the basis of past experience and available scientific information, range managers are often able to specify a safe minimum standard,[4] and all the states of the rangeland below the safe minimum standard can be thought of as crisis states. Next, it is necessary to put in place a well designed plan of action. In the context of a particular rangeland, this might mean the use of a specialized grazing system such as short duration grazing.[5]

As discussed in the previous two paragraphs, even with a well designed plan of action in place, it is still *possible* that a managed rangeland will hit a crisis state. Accordingly, a key question now relates to the length of time until the crisis state is reached. Put differently, the range manager would like to know how long it takes for the rangeland to hit a crisis state. In addition, what does the answer to this query depend on? Finally, is the answer history dependent or independent? In other words, does the answer to the length of time until crisis query depend on the state in which the range manager's plan of action is put in place? Obviously, answers to these queries are critical for successful rangeland management. Even so, to the best of our knowledge, there are no previous analyses of these queries. Therefore, in this chapter, we provide a theoretical analysis of these hitherto unanalyzed issues.

The rest of this chapter is organized as follows. Section 2.1 describes our discrete-time Markov chain theoretic model[6] of an arbitrary rangeland. Section 2.2 provides an answer to the how long until crisis query for a range management regime that we call slack. Section 2.3 answers the same query for a range management regime that we term stern. Section 2.4 uses a numerical example to illustrate the methods used in the previous two sections. Section 3 concludes and discusses potential extensions of this chapter's research.

2. The Length of Time until Crisis

2.1. The Theoretical Framework[7]

A stochastic process $\{Z_t\}$ is said to be a Markov process if it possesses the attribute that, given the value of Z_t, the values of Z_u for $u > t$ are not impacted by the values of Z_s for $s < t$. In words, the likelihood of any specific future behavior of the process, when its present state is known, is not altered by additional knowledge relating to its past behavior. A discrete-time Markov chain is a Markov process whose state space is either finite or countable and whose time index set is given by $I = (0,1,2,3,...)$. In symbols (see Taylor and Karlin (1998, p. 95)), the Markov property tells us that

[4] For more on the practical aspects of range management under uncertainty, see Campbell and Stafford Smith (2000). For a general discussion of the use of the safe minimum standard in the presence of potential irreversibilities, see Farmer and Randall (1998).

[5] For more on short duration grazing, see Batabyal (2001, 2002, also chapter 9), Holechek *et al.* (2001), and Batabyal and Yoo (2002, also chapter 5).

[6] Textbook accounts of discrete-time Markov chains can be found in Ross (1996, 2000) and in Taylor and Karlin (1998).

[7] This theoretical framework is adapted from Batabyal and Beladi (2002).

$$Prob\{Z_{n+1} = j / Z_0 = i_0, ..., Z_{n-1} = i_{n+1}, Z_n = i\} = Prob\{Z_{n+1} = j / Z_n = i\}, \quad (1)$$

for all time points n and all states $i_0, ..., i_{n-1}, i, j$. Further, the one-step transition probabilities of a discrete-time Markov chain are given by $P_{ij}^{n,n+1} = Prob\{Z_{n+1} = j / Z_n = i\}$. Finally, the reader should note that a discrete-time Markov chain spends one unit of time in a state before making a transition to some state.

Now consider an arbitrary dynamic and stochastic rangeland. Formally, we shall think of this rangeland as a discrete-time Markov chain with a finite number of rank ordered states *0,1,2,...,S.* This means that the states toward the bottom of the rank order are undesirable states and the states toward the top of the rank order are the desirable states. In particular, state 0 is the least desirable state and state S is the most desirable state. To fix ideas, the reader may want to think of state S as one in which there is no human use of the rangeland.

As indicated in section 1, our rangeland will generally have a number of undesirable states but not all of them will be crisis states. We hypothesize that there is a single crisis state and we shall call this state, state 0. We now have to specify the state in which our range manager inherits the rangeland under study. In principle, this could be any state *i,* where $i = 1, ..., S$. Nevertheless, in what follows, to place the crisis issue in sharp focus, we shall study the situation that is as propitious as possible for our range manager. This most propitious situation arises when the manager inherits the rangeland in state *S.* Now, given a particular range management regime, we are interested in determining the expected value of *T,* where T is the time until our rangeland hits the crisis state 0. In technical language, we say that state 0 is the absorbing state and T is the time to absorption for our discrete-time Markov chain. In symbols, we are interested in ascertaining c_i where $c_i = E[T / Z_0 = i]$. Note that i in this conditional expectation expression can, in general, be any state from *1,2,...,S.* However, as explicated previously, we shall conduct the analysis for the case in which the rangeland is initially in state *S.* This completes the discussion of our theoretical framework. We now discuss the length of time until crisis issue in the context of first, a slack range management regime, and then, a stern range management regime.

2.2. The Slack Regime

Let us first provide a stochastic delineation of the slack management regime. To this end, suppose that our rangeland is in state S in time period *n.* Then, in time period $n+1$, the position of the rangeland is uniformly distributed over the relevant states $0,1,2,...,S-1$. As indicated in section 2.1, the manager inherits the rangeland, and hence our analysis begins, in state *S.* This management regime is slack in the sense that even though the manager inherits the rangeland in the best possible state *S,* his or her managerial actions are expendable and hence they are largely futile in keeping the rangeland away from the undesirable states in general and the crisis state in particular. Consequently, given that the rangeland is now in state *S,* it is just as likely that the rangeland will next be in state 0 as it is that it will next be in state $S-3$.

Examples of such slack range management regimes exist in many parts of the world. For instance, in the western United States, Heady and Child (1994, pp. 6-7) have noted that until recently, management has been such that "high livestock numbers, uncontrolled grazing on public lands, and lack of knowledge or care for the land mainly caused the destruction [of rangelands]." In Sukhomajri, a village in the north Indian state of Haryana, traditional cattle rearers have often adopted practices that have "accentuated the degradation of forest and pasture land" (Chopra and Rao, 1997, p. 311). Similarly, in the context of countries such as Iraq, Jordan, and Syria, Tutwiler and Bailey (1997, p. 285) have pointed out that livestock management practices are such that "early grazing and overgrazing are common. Palatable grazing species are not given enough time to regenerate from one year to the next, and the quality of pasture is deteriorating." Finally, in Africa, the range management practices of some pastoralists have been "widely blamed for overstocking, trampling of soils around water points..., destruction of woodland through burning..., and prevention of organic buildup in the soil by feeding crop residues to animals" (Webb and Coppock, 1997, p. 250).

We now compute the average length of time until crisis for our slack range management regime. We shall use the technique of first step analysis. To this end, recall that states $1,2,...,S,$ are the transient states of our rangeland. Consequently, our principal task now is to compute the inverse of what is called the $(I - Q)$ matrix for the above specified transient states (see Taylor and Karlin, 1998, p. 172). Let us call this inverse matrix W. In other words $W = (I - Q)^{-1}$. Now observe that each of the relevant states is visited by our rangeland at most once and hence the $(i, j)th$ element of the W matrix is given by

$$W_{ij} = Prob\{ever\ visit\ state\ j/Z_0 = i\}. \tag{2}$$

Some reflection and equation (1) tell us that $W_{ii} = 1$ and that $W_{ij} = 0$ for $j > i$. Further, we can also infer that

$$W_{i,i-1} = Prob_{i,i-1} = \frac{1}{i}, \tag{3}$$

that

$$W_{i,i-2} = Prob_{i,i-2} + Prob_{i,i-1}W_{i-1,i-2} = \frac{1}{i} + \frac{1}{i}(\frac{1}{i-1}) = \frac{1}{i-1}, \tag{4}$$

and that

$$W_{i,i-3} = Prob_{i,i-3} + Prob_{i,i-2}W_{i-2,i-3} + Prob_{i,i-1}W_{i-1,i-3} = \frac{1}{i}(1 + \frac{1}{i-2} + \frac{1}{i-2}) = \frac{1}{i-2}. \tag{5}$$

Continuing in the above described manner, we deduce that

$$W_{ij} = \frac{1}{j+1}, 1 \leq j < i \tag{6}$$

Recall that our objective is to determine $c_S = E[T / Z_0 = S]$ the average time taken to hit the crisis state, given that our manager inherits the rangeland in the most desirable state *S*. To compute this expectation we shall use equation 7.13 in Taylor and Karlin (1998, p. 172). This equation tells us that

$$c_S = E[T / Z_0 = S] = \sum_{j=1}^{j=S} W_{Sj} = W_{S1} + W_{S2} + W_{S3} + ... + W_{SS}. \tag{7}$$

Because $W_{SS} = 1$, equation (7) can be simplified to

$$c_S = E[T / Z_0 = S] = \sum_{j=1}^{j=S} W_{Sj} = 1 + \sum_{j=1}^{j=S-1} \frac{1}{j+1} \tag{8}$$

Now simplifying the second summation on the right-hand-side (RHS) of equation (8), we get

$$c_S = E[T / Z_0 = S] = 1 + \frac{1}{2} + \frac{1}{3} + ... + \frac{1}{S} = \sum_{b=1}^{b=S} \frac{1}{b} \tag{9}$$

Equation (9) gives us the answer that we seek. In words, this equation tells us that for the slack management regime, the average time until crisis is given by the summation of the reciprocal of the finite states, the summation being carried out from states from 1 to *S*. For the asymptotic case in which $S \rightarrow \infty$, the term $\sum_{b=1}^{b=S} \frac{1}{b}$ is well approximated by the natural logarithm function. The use of this approximation gives us a more compact answer to the how long until crisis query. In particular, we get

$$c_S = E[T / Z_0 = S] = \log_e(S) \tag{10}$$

In words, equation (10) tells us that the expected amount of time until our slackly managed rangeland hits the crisis state is given by the logarithm of the state in which the manager inherits the rangeland.

As noted in Holechek *et al.* (2001, p. 185), it is common for range managers to think of the condition of a rangeland in terms of four states called excellent, good, fair, and poor. Let us rank order these four states so that state 0 is the poor or crisis state, state 1 is the fair state, state 2 is the good state and state 3 is the excellent state. Now assume that our slack manager inherits the rangeland in the excellent state. Then, equation (9) tells us that it will take, on average, 1.83 time periods for the rangeland to reach the crisis state.

Our analysis thus far permits us to draw an important conclusion concerning rangeland management. Equations (9) and (10) clearly tells us that the answer to our how long until crisis query depends on state S, i.e., on the state in which our range manager inherits the rangeland. Therefore, from a management perspective, it is significant to understand that the initial condition matters. More generally, given our rank ordering of the states of the rangeland, the more desirable the state in which the manager inherits the rangeland, the longer it will take for the rangeland to hit the crisis state. In contrast, if the initial state of inheritance is comparatively undesirable, then it will take much less time for the rangeland to hit the crisis state. We now proceed to discuss the answer to the how long until crisis query when the rangeland under study is managed by a stern manager.

2.3. The Stern Regime

Once again, we begin with a probabilistic delineation of the stern range management regime. Suppose that our rangeland is in state j in time period n. Then, in time period $n+1$, the position of the rangeland is state 0 (the crisis state) with probability $1/j$, and its position is state k, where k is any one of the states $1,2,3,...j-1$, with probability $2k/j^2$. As in section 2.2, our range manager inherits the rangeland, and therefore our analysis begins, in state S. The present range management regime is stern in two senses. First, given that the rangeland is now in state S, the probability of hitting the crisis state next is *not* identical to the likelihood of hitting some other (non-crisis) state. Second, the probability of hitting the crisis state next depends on where the rangeland is initially. In particular, the more desirable the initial state, i.e., the closer the initial state is (in rank order terms) to state S, the less likely it is that the rangeland will hit the crisis state.

There are many range management regimes that are stern in the two senses in which we're using the term stern in this chapter. A specific example is the management of pastoral lands in Ngorongoro, Tanzania (see Homewood and Rodgers (1991)). More generally, all over Africa, including the arid and the semiarid parts of this continent, instances of what we have called stern range management have been documented.[8] This has led Agrawal (1998) and others to note that pastoralists are not always irrational or inefficient users of rangelands. Indeed, in this context, it has been pointed out that at least some pastoralists are appropriately viewed as "shrewd managers of livestock who can uses resources with a sensitive recognition of the capacities of their environments" (Agrawal, 1998, p. 32).

Let us now compute the expected time until crisis for our stern range management regime. As in section 2.2, we shall begin by computing the W matrix. In this regard, note that also as in section 2.2, equation (2) holds and once again we have $W_{ii}=1$ and $W_{ij}=0$ for $j>i$. So, our next task is to determine W_{kj} for $1\le j<k$. Put differently, the outstanding task is to find the solution to the following first step equation in W_{kj}

[8] See Anderson and Grove (1987), Bonte and Galaty (1991), and Behnke *et al.* (1993) for a more detailed corroboration of this claim.

$$W_{kj} = Prob_{kj} + \sum_{l=j+1}^{l=k-1} Prob_{kl} W_{lj} . \tag{11}$$

We hypothesize that

$$W_{kj} = 2\{\frac{k+1}{k}\}\{\frac{j}{(j+1)(j+2)}\}, \quad 1 \le j < k \tag{12}$$

solves equation (11). To demonstrate that the expression in (12) does indeed solve equation (11), let us simplify the RHS of equation (11). We get

$$\text{RHS} = \frac{2j}{k^2} + \sum_{l=j+1}^{l=k-1} \frac{2l}{k^2} 2\{\frac{l+1}{l}\}\{\frac{j}{(j+1)(j+2)}\} = \frac{2j}{k^2} + \{\frac{2j}{k^2}\}\{\frac{2}{(j+1)(j+2)}\}\sum_{l=j+1}^{l=k-1}(l+1) \tag{13}$$

$$= \frac{2j}{k^2} + \{\frac{2j}{k^2}\}\{\frac{2}{(j+1)(j+2)}\}\{\frac{k(k+1)}{2} - \frac{(j+1)(j+2)}{2}\} \tag{14}$$

$$= 2\{\frac{k+1}{k}\}\{\frac{j}{(j+1)(j+2)}\} = W_{kj}, \tag{15}$$

which is what we wanted to show (see equation (12)).

Having computed the W matrix, we now want to calculate the expected time until crisis when the stern management regime is in place. Mathematically, we want to calculate $c_S = E[T / Z_0 = S]$ Using the section 2.2 logic, we get

$$c_S = E[T / Z_0 = S] = \sum_{j=1}^{j=S} W_{Sj} = 1 + 2\{\frac{S+1}{S}\}\sum_{j=1}^{j=S-1} \frac{j}{(j+1)(j+2)} \tag{16}$$

In the case of a finite number of states, the answer to our query given in equation (16) can be expressed in an alternate form. Expressing the answer in this alternate form, we get

$$c_S = E[T / Z_0 = S] = 2\{\frac{S+1}{S}\}\{1 + \frac{1}{2} + \frac{1}{3} + \ldots + \frac{1}{S}\} - 3. \tag{17}$$

Equation (17) tells us that for the stern range management regime, the expected time until crisis is given by the product of two terms less the constant 3. The first term in this product is the ratio $2(S+1)/S$ and the second term is the summation of the reciprocal of the finite states, where the summation is carried out from states from 1 to S. Comparing equations (9) and (17), we see that this second term in the product on the RHS of equation (17) is the same as the solution that we obtained to the how long until crisis query for the slack range management regime. This tells us that the two answers to the how long until crisis query are obviously related to each other.

Now, for the two range management regimes that we have studied in this chapter, suppose that the number of states of the rangeland is the same, i.e., S, and that the manager inherits the rangeland in the same state, i.e., in state S. In this case, we would expect the time taken until crisis in the stern range management regime to exceed the time taken until crisis in the slack range management regime. A quick check with the four state range example of section 2.2 tells us that this is indeed the case. Specifically, it takes 1.83 time periods for the rangeland to hit the crisis state in the slack range management regime, and — using equation (17) we see that — it takes 1.89 time periods for the rangeland to hit the crisis state in the stern range management regime.

What happens in the stern range management regime when the number of states of the rangeland approaches infinity? Taking the limit as $S \rightarrow \infty$ in equation (17), we get

$$c_S = E[T / Z_0 = S] = 2\log_e(S) - 3 \tag{18}$$

In other words, the expected amount of time until our sternly managed rangeland hits the crisis state is given by twice the logarithm of the state in which the manager inherits the rangeland less the constant 3. Comparing the equation (18) result with the corresponding result in equation (10), we see that even in the asymptotic case, it generally takes longer for the rangeland to hit the crisis state with the stern range management regime. As a quick check, when S=1,000,000, using equations (10) and (18) it is straightforward to confirm that it takes approximately 14 time periods to hit the crisis state with the slack range management regime and it takes about 25 time periods to hit the crisis state with the stern range management regime.

As far as the management of dynamic and stochastic rangelands is concerned, equation (18) clearly tells us that the answer to our how long until crisis query, once again, depends on the state (S) in which the manager inherits the rangeland. Therefore, from a range management perspective, as in section 2.2, the initial condition matters. More generally, given our rank ordering of the states of the rangeland, if the initial state of inheritance is comparatively undesirable, then it will take a small amount of time for the rangeland to hit the crisis state. In contrast, the more desirable the state in which the manager inherits the rangeland, the longer it will take for the rangeland to hit the crisis state. We now provide a numerical example to show how the average time until crisis can be computed when the transition probabilities of the discrete-time Markov chain are known to the range manager.

2.4. A Numerical Example

As in sections 2.2 and 2.3, consider a rangeland with four states called Poor, Fair, Good, and Excellent. Reversing the rank ordering employed in the previous two sections, we shall say that state 0 is the Excellent state, state 1 is the Good state, state 2 is the Fair state, and state 3 is the Poor state. In addition, we reckon that the transition probability matrix of this four state rangeland is

$$P = \begin{bmatrix} P_{00} & P_{01} & P_{02} & P_{03} \\ P_{10} & P_{11} & P_{12} & P_{13} \\ P_{20} & P_{21} & P_{22} & P_{23} \\ P_{30} & P_{31} & P_{32} & P_{33} \end{bmatrix} = \begin{bmatrix} 0.4 & 0.3 & 0.2 & 0.1 \\ 0 & 0.7 & 0.2 & 0.1 \\ 0 & 0 & 0.9 & 0.1 \\ 0 & 0 & 0 & 1 \end{bmatrix} \quad (19)$$

Scrutinizing the above transition probability matrix, it is clear that states 0, 1, and 2 are transient states. However, because $P_{33} = 1$ state 3, the so called Poor state, is the only crisis state. Now, consistent with the discourse in the previous two sections of this chapter, suppose that our manager inherits this rangeland in state 0, the most desirable or Excellent state. Then, given that the initial state is 0, the first row of the P matrix above is germane. Specifically, we see that as far as transitions are concerned, there are three possibilities. The rangeland may next hit the crisis state 3 in which case the expected time that we're after is 1 period long. However, the rangeland may stay in state 0 or it may next visit states 1 or 2. Therefore, to compute the mean time until crisis, we have to weight these various contingencies with the germane probabilities that these contingencies will in fact occur. Doing so (see Taylor and Karlin, 1998, pp. 116-123 for the methodology), we see that the relevant average time that we seek is 10. In other words, in this example, the answer to the how long until crisis query is that it will take 10 time periods for the rangeland to hit the crisis state 3.

3. Conclusions

In the context of rangeland management over time and under uncertainty, we provided a theoretical perspective on the hitherto unstudied how long until crisis query. In particular, we studied slack and stern range management regimes and we obtained answers to the expected time until crisis query for both the finite S and the large S cases. In addition, we showed that the mean time until crisis is shorter in the slack range management regime and longer in the stern range management regime. Finally, we pointed out that the answer to the how long until crisis query depends on the initial state. This is the reason for saying that the initial condition matters in rangeland management.

The analysis of this chapter can be extended in a number of different directions. In what follows, we suggest two possible extensions. First, we studied models in which there is a single crisis state. Accordingly, an obvious way to generalize the analysis of this chapter would be to study the how long until crisis query with models in which there are several crisis states. Second, we used a discrete-time Markov chain to model and then examine the slack and the stern range management regimes. A weakness of this approach is that the time spent in a state before making a transition to some other state is non-random and always one unit long. The reasonableness of this assumption is open to question. Hence, it would be expedient to eschew this assumption and research the how long until crisis query with, for instance, a semi-Markov process. Studies that integrate these facets of the problem into the analysis will increase our understanding of the complexities of rangeland management over time and under uncertainty.

REFERENCES

Agrawal, A. 1998. *Greener Pastures*. Duke University Press, Durham, NC.

Anderson, D., and Grove, R. Eds. 1987. *Conservation in Africa*. Cambridge University Press, Cambridge, UK.

Batabyal, A.A. 2001. Some Theoretical Aspects of Short Duration Grazing, *International Journal of Ecology and Environmental Sciences*, 27, 215-220.

_____ . 2002. An Upper Bound for the Transient Resilience of a Rangeland that is Managed Using Cell Grazing, *Journal of Economic Research*, 7, 151-160.

_____ . 2003. A Note on First Step Analysis and Rangeland Management Under Uncertainty, *Unpublished Manuscript*, Rochester Institute of Technology.

Batabyal, A.A., and Godfrey, E.B. 2002. Rangeland Management Under Uncertainty: A Conceptual Approach, *Journal of Range Management*, 55, 12-15.

Batabyal, A.A., and Yoo, S.J. 2002. The Steady State Distribution of Animals in Short Duration Grazing, *Applied Economics Letters*, 9, 695-698.

Batabyal, A.A., and Beladi, H. 2002. Renewable Resource Management in Developing Countries: How Long Until Crisis? Forthcoming, *Review of Development Economics*.

Behnke, R.H., Scoones, I., and Kerven, C. Eds. 1993. *Range Ecology at Disequilibrium*. Overseas Development Institute, London, UK.

Bonte, P., and Galaty, J. 1991. Introduction, in P. Bonte and J. Galaty, (Eds.), *Herders, Warriors, and Traders*. Westview Press, Boulder, CO.

Campbell, B.D., and Stafford Smith, D.M. 2000. A synthesis of Recent Global Change Research on Pasture and Rangeland Production: Reduced Uncertainties and Their Management Implications, *Agriculture, Ecosystems, and Environment*, 82, 39-55.

Center for Science and Environment. 1990. *Human-Nature Interactions in a Central Himalayan Village: A Case Study of Village Bembru*. Center for Science and Environment, New Delhi, India.

Chopra, K., and Rao, C.H.H. 1997. Institutional and Technological Perspectives on the Links Between Agricultural Sustainability and Poverty: Illustrations from India, in S.A. Vosti and T. Reardon, (eds.), *Sustainability, Growth, and Poverty Alleviation*. Johns Hopkins University Press, Baltimore, MD.

Dasgupta, P. 1996. The Economics of the Environment, *Environment and Development Economics*, 1, 387-428.

Dasgupta, P., and Maler, K.G. 1997. The Resource-Basis of Production and Consumption: An Economic Analysis, in P. Dasgupta and K.G. Maler, (Eds.), *The Environment and Emerging Development Issues*, Volume 1. Clarendon Press Oxford, Oxford, UK.

Farmer, M.C., and Randall, A. 1998. The Rationality of a Safe Minimum Standard, *Land Economics*, 74, 287-302.

Heady, H.F., and Heady, E.B. 1982. *Range and Wildlife Management in the Tropics*. Longman Inc., New York, NY.

Heady, H.F., and Child, R.D. 1994. *Rangeland Ecology and Management*. Westview Press, Boulder, CO.

Ho, P. 2000. China's Rangelands Under Stress: A Comparative Study of the Pasture Commons in the Ninxia Hui Autonomous Region, *Development and Change*, 31, 385-412.

Homewood, K., and Rodgers, W. 1991. *Maasailand Ecology*. Cambridge University Press, Cambridge, UK.

Holechek, J.L., Pieper, R.D., and Herbel, C.H. 2001. *Range Management*, 4th edition. Prentice Hall, Upper Saddle River, NJ.

Jodha, N.S. 1980. The Process of Desertification and the Choice of Interventions, *Economic and Political Weekly*, 15, 1351-1356.

Ross, S.M. 1996. *Stochastic Processes*, 2nd edition. Wiley, New York, NY.

_____ . 2000. *Introduction to Probability Models*, 7th edition. Harcourt Academic Press, San Diego, CA.

Sandford, S. 1983. *Management of Pastoral Development in the Third World*. Wiley, New York, NY.

Taylor, H.M., and Karlin, S. 1998. *An Introduction to Stochastic Modeling*, 3rd edition. Academic Press, San Diego, CA.

Tutwiler, R.N., and Bailey, E. 1997. Agricultural Growth and Sustainability: Conditions for their Compatibility in the Rainfed Production Systems of West Asia and North Africa, in S.A. Vosti and T. Reardon, (eds.), *Sustainability, Growth, and Poverty Alleviation*. Johns Hopkins University Press, Baltimore, MD.

Webb, P., and Coppock, D.L. 1997. Prospects for Pastoralism in Semi-Arid Africa, in S.A. Vosti and T. Reardon, (eds.), *Sustainability, Growth, and Poverty Alleviation*. Johns Hopkins University Press, Baltimore, MD.

Chapter 7

RANGELAND MANAGEMENT UNDER UNCERTAINTY: A CONCEPTUAL APPROACH

with E. B. Godfrey

ABSTRACT

A conceptual approach is commonly needed to provide guidance for empirical studies concerning the use of renewable resources such as rangelands. The theoretical model constructed in this chapter captures the essential aspects of dynamic and stochastic issues associated with the management of rangelands. We use our approach to discuss the connections between this chapter's model and range policy. Specifically, we point out scenarios in which there is limited or no role for policy. This is compared to scenarios where policy has a significant role to play in ensuring the sustainable use of rangelands. Finally, we suggest two ways in which our approach might be extended and used in an empirical application.

Keywords: Dynamic, Policy, Resource, Stochastic

1. INTRODUCTION

Rangelands are an important renewable resource. In addition to performing a number of salient ecological functions, rangelands provide humans with consumable products such as red meat, fiber and water, and non-consumptive services such as recreation and wildlife viewing. As such, it is no surprise that range managers have systematically attempted to manipulate "range components to obtain the optimum combination of goods and services for society on a sustained basis" (Holechek *et al.*, 2001, p. 5). The management of rangelands is not easy because range managers rarely have complete information about the impact of actions that may be taken. This state of affairs is in part due to a lacuna in the range management literature. Although previous studies (see Lambert and Harris (1990), Passmore and Brown (1991), McCluskey and Rausser (1999), Batabyal (2000), and Batabyal *et al.* (2001)) have studied aspects of range management in the presence of uncertainty, there are very few studies that have explicitly modeled the *connections* between uncertainty, the ecology, and the management of rangelands over time.

Given this state of affairs, a conceptual approach is needed to provide guidance for empirical studies concerning policies that affect the use of rangelands. The model outlined below has three desirable features. First, the approach is consistent with an important ecological model of range behavior, namely, the state-and-transition model of Westoby *et al.* (1989). Second, the approach expressly accounts for the role that uncertainty plays in the temporal evolution of managed rangelands. Finally, by exploring the connections between policy and range behavior, our approach makes transparent the role that policy can play, and, on occasion not play, in ensuring the sustainable use of rangelands.

The rest of this chapter is organized as follows. The formal state-and-transition model is outlined in section 2. This model provides the basis for a discussion of the policy implications associated with the dynamic and the stochastic structure of rangelands. This discussion is followed by an extension of the model in two ways that are likely to be important and that are not captured in the basic model. Section 3 concludes and outlines areas for future research.

2. A "State-and-Transition" Model of a Rangeland

2.1. Preliminaries

As discussed but not formalized in Westoby *et al.* (1989), rangeland dynamics can be usefully described in terms of a discrete number of states and a set of transitions between these states. Range managers often use the term "condition class" to refer to these states. Moreover, as noted in Holechek *et al.* (2001, p. 185), it is common to think of rangelands as existing in one of four possible condition classes such as excellent, good, fair, and poor. Consequently, we assume that the rangeland under study can exist in four possible states, but the number of states could easily be extended, without any loss of generality, to any countable number of states.

Environmental factors such as fire and/or actions such as the alteration of the stocking rate result in the rangeland making transitions from one condition class to the other. To formalize this notion, suppose that our rangeland makes state transitions in accordance with a discrete-time Markov chain (see Ross (1997, pp 157-234) and Perrings (1998)). In other words, when in state $i, i = 1,...,4,$ the rangeland will make a transition to state $j, j = 1,...,4,$ with probability P_{ij}. From a management perspective, not all condition classes are equally desirable. In particular, although other combinations are possible and analyzable, it seems reasonable to think of excellent (state 1) and good (state 2) as constituting the desirable set of states, and fair (state 3) and poor (state 4) constituting the undesirable set of states of the rangeland. Formally, we have $D = \{Excellent\,(state\ 1), Good\,(state\ 2)\}$ and $U = \{Fair\,(state\ 3), Poor\,(state\ 4)\}$.

The aim of our range manager is to use the policies at his disposal (changing the stocking rate, burning, introducing plant populations) to keep the rangeland in the desirable set D for as long as possible. However, because this manager cannot be certain about the appropriateness of a particular policy (say, the positive and negative implications of a burn) and because state transitions occur not only because of managerial actions but also because of environmental factors, it is possible that despite the manager's best intentions, the rangeland will end up in a state in the undesirable set U. As such, the manager would like to know two things. First, what is the expected amount of time $(\overline{U})$ that the rangeland spends in set U, given that it is currently in a state in U. Second, what is the expected amount of time $(\overline{D})$ the rangeland spends in set D, given that it is presently in a state in D. Let us now compute these two expectations.

2.2. Two Expectations

Let $\pi_k, k = 1,...,4$, denote the stationary probabilities (see Ross (1997, pp. 172-182)) of the rangeland. Now, for state $i \in D$ and state $j \in U$, the rate at which the rangeland enters state j from state i is $\pi_i P_{ij}$. From this it follows that the rate at which our rangeland enters state j from a state in the desirable set D is $\sum_{i=1}^{i=2} \pi_i P_{ij}$. With these two pieces of information, we conclude that the rate at which our rangeland moves from the desirable to the undesirable set of states is $\sum_{j=3}^{j=4} \sum_{i=1}^{i=2} \pi_i P_{ij}$. Similarly, but now in terms of the two expectations $\overline{D}$ and $\overline{U}$, the rate at which the rangeland moves from the desirable to the undesirable set of states is $1/(\overline{D}+\overline{U})$. Equating these last two expressions, we get

$$\sum_{j=3}^{j=4} \sum_{i=1}^{i=2} \pi_i P_{ij} = \frac{1}{\overline{D}+\overline{U}}. \tag{1}$$

To compute $\overline{D}$ and $\overline{U}$ explicitly, we need a second equation linking these two expectations. This equation can be obtained by noting two facts. First, the percentage of time that the rangeland is in the desirable set of states is $\sum_{i=1}^{i=2} \pi_i$ Second, the proportion of time the rangeland spends in the desirable set of states is $\overline{D}/(\overline{D}+\overline{U})$. Combining these two facts, we get

$$\sum_{i=1}^{i=2} \pi_i = \frac{\overline{D}}{\overline{D}+\overline{U}}. \tag{2}$$

Now using equations (1) and (2), it is straightforward to verify that

$$\overline{D} = \frac{\sum_{i=1}^{i=2} \pi_i}{\sum_{j=3}^{j=4} \sum_{i=`}^{i=2} \pi_i P_{ij}} \text{ and } \overline{U} = \frac{\sum_{j=3}^{j=4} \pi_j}{\sum_{j=3}^{j=4} \sum_{i=`}^{i=2} \pi_i P_{ij}} \tag{3}$$

Inspecting equation (3) and recalling the properties of the stationary probabilities of a Markov chain (see Ross (1997, pp. 172-182)) it is clear that although managerial actions can influence the expected amount of time that the rangeland spends in the desirable and the undesirable set of states, ultimately, it is the transition probabilities, i.e., the $P_{ij}' s$ that should be the target of range policy. To see this clearly, consider the connections between alternate policies and the transition probability matrix, P, of our four state rangeland (Markov chain). P can be written as

$$P = \begin{bmatrix} P_{11} & P_{12} & P_{13} & P_{14} \\ P_{21} & P_{22} & P_{23} & P_{24} \\ P_{31} & P_{32} & P_{33} & P_{34} \\ P_{41} & P_{42} & P_{43} & P_{44} \end{bmatrix} \tag{4}$$

In this matrix, the first row and column denote the excellent state (state l), the second row and column denote the good state (state 2), and so on. Thus, the probability of making a transition to the excellent state (state l), given that the rangeland is currently in the poor state (state 4), is denoted by P_{41}. The other elements of the matrix P have similar interpretations. Now, although a range manager may have disparate policy goals, as indicated in the last paragraph of section 2.1, one reasonable goal is to ensure that the rangeland stays in the desirable set of states D, for as long as possible. This can be done by influencing the transition probabilities in the P matrix. Inspection of equation (4) and some reflection should enable the reader to see that from the standpoint of range policy, the matrix P can be usefully partitioned into four zones. These four zones correspond to the following four sub-matrices

$$P_1 = \begin{bmatrix} P_{11} & P_{12} \\ P_{21} & P_{22} \end{bmatrix}, \quad P_2 = \begin{bmatrix} P_{13} & P_{14} \\ P_{23} & P_{24} \end{bmatrix}, \quad P_3 = \begin{bmatrix} P_{31} & P_{32} \\ P_{41} & P_{42} \end{bmatrix}, \textit{and } P_4 = \begin{bmatrix} P_{33} & P_{34} \\ P_{43} & P_{44} \end{bmatrix}. \tag{5}$$

If the rangeland is in P_1 then it is in the desirable set of states (1 or 2). As such, this is a zone in which no major policy initiatives are required to ensure the well-being of this rangeland. In contrast, when the rangeland is in P_4 it is in the undesirable set of states (3 or 4) and the probability of bringing it to the desirable set is zero. This means that in this zone, there is no role for policy. Put differently, the rangeland is so degraded that it cannot possibly be restored to one of the desirable states. The two intermediate zones corresponding to P_2 and P_3 are of particular interest. This is where there is the greatest scope for policy. In other words, in these two zones, resource management can, *inter alia*, change the stocking rate, remove noxious plants, keep livestock away from streamside areas, and introduce plant populations, in order to ameliorate the rangeland. Inspecting P_2, we see that the probabilities here involve moving from the desirable to the undesirable set of states. As such, it is clear that when in this zone, the objective of policy should be to take those actions that will minimize the probabilities in this sub-matrix and improve the rangeland resource. In contrast, inspection of P_3 tells us that the probabilities here involve moving from the undesirable to the desirable set of states. This means that when in this zone, policy should confine itself to those actions that maximize the probabilities in this sub-matrix and thereby improve the range. This way of looking at the task of range management provides a clear indication as to when there is a substantial role and when there is a very limited role for managerial actions in ensuring the sustainable use of our rangeland. We now illustrate this conceptual approach with a numerical example.

2.3. An Example

Suppose that the transition probability matrix in equation (4) is

$$P = \begin{bmatrix} 1/2 & 1/2 & 0 & 0 \\ 1/4 & 1/4 & 1/4 & 1/4 \\ 1/5 & 1/5 & 2/5 & 1/5 \\ 0 & 0 & 1/4 & 3/4 \end{bmatrix} \tag{6}$$

While the transition probabilities in equation (6) are illustrative, subject to data availability, they can be estimated using econometric techniques (Ethridge *et al.* (1985)). In a practical setting, these probabilities are useful because they provide rangeland managers with information about how the interaction of managerial actions and environmental factors affect range condition. Let us compute $\overline{U}$ and $\overline{D}$. To perform this computation, we shall use a three-step procedure. First, we observe that the stationary probabilities satisfy

$$\pi_1 = (1/2)\pi_1 + (1/4)\pi_2 + (1/5)\pi_3,$$

$$\pi_2 = (1/2)\pi_1 + (1/4)\pi_2 + (1/5)\pi_3,$$

$$\pi_3 = (1/4)\pi_2 + (2/5)\pi_3 + (1/4)\pi_4$$

and $\pi_4 = 1 - \sum_{i=1}^{i=3} \pi_1$ (see Ross (1997, p. 173) for additional details). These four equations can be solved to yield $\pi_1 = 4/21$, $\pi_2 = 4/21$, $\pi_3 = 5/21$ and $\pi_4 = 8/21$. Second, we use these stationary probabilities and the transition probabilities in equation (6) to determine the rate at which our rangeland moves from the desirable to the undesirable set of states, i.e., the denominator of the two expressions for $\overline{D}$ and $\overline{U}$ in equation (3). This rate equals 2/21. Finally, we compute $\overline{D}$ and $\overline{U}$ by using this rate (2/21), and the expressions in equations (3) and (6). We get $\overline{D} = 4$ and $\overline{U} = 6.5$.

What are the implications of these computations for range policy? These computations tell us that, on average, our rangeland moves from the desirable to the undesirable set of states 2/21 or approximately 10% of the time. The rangeland resource stays in the undesirable set for 6.5 units of time on average. This is followed by a time span, on average four units long, during which the rangeland is in the desirable set of states. This kind of rangeland behavior is entirely consistent with the state-and-transition model of Westoby *et al.* (1989). Further, note that in ecology, the stability concept known as persistence refers to "how long a [resource] lasts before it is changed to a new value" Pimm (1991, p. 14). In the context of this chapter, the expectations $\overline{D}$ and $\overline{U}$ can be interpreted as the persistence of, respectively, the desirable and the undesirable set of states. This means that when range use policy is directed to the maximization (minimization) of $\overline{D}(\overline{U})$, society benefits in an economic and in an ecological sense.

2.4. Two Extensions

The discrete-time Markov chain model described above nicely captures the essential elements of dynamic and stochastic rangelands. However, the scope of this model is restricted by the presence of two features, First, range policy cannot alter the transition probabilities (the P_{ij}' s) because these probabilities are stationary. Second, range policy also cannot change the amount of time the rangeland spends in a particular state because the discrete-time Markov chain spends one unit of time in each state before making a transition. How might we account for these two features in our model? We now answer this question.

First, let us consider the case of non-stationary transition probabilities. In this case, the probability of making a transition from condition class i at time $t-1$ to condition class j at time t depends on t. To formalize this time dependence, we write $P_{ij}^{t-1,t}$ instead of P_{ij}. Similarly, the transition probability matrix in equation (4) will now have to be replaced by a sequence of transition matrices. In other words, instead of working with the single matrix P of equation (4), we now work with a sequence of matrices $[P_t]_{t=1}^{\infty}$. Once we specify an initial vector that gives a probability distribution over the states of the non-stationary Markov chain, we have completely described our rangeland. Then we can investigate the limiting behavior of the rangeland under study as time approaches infinity. This rangeland may or may not converge to a limiting vector. For instance, consider the two-condition class rangeland with transition matrices

$$P_{2t-1} = \begin{bmatrix} 1-\dfrac{1}{2t-1} & \dfrac{1}{2t-1} \\ 1-\dfrac{1}{2t-1} & \dfrac{1}{2t-1} \end{bmatrix},\ P_{2t} = \begin{bmatrix} \dfrac{1}{2t} & 1-\dfrac{1}{2t} \\ \dfrac{1}{2t} & 1-\dfrac{1}{2t} \end{bmatrix},\ t = 1,2,3,... \quad (7)$$

In this case, it can be shown, using the methods outlined by Isaacson and Madsen (1976, pp. 140-141) that this non-stationary Markov chain displays a kind of limiting behavior known as weak ergodicity.

We now address the "unit time in a state before transition" feature. Ideally, we would like rangeland management policy to be conducted in a way that the rangeland resource under study is in the desirable (undesirable) set of states for relatively long (short) periods of time. One way to model this is as follows. We let the transition probabilities be stationary. However, instead of having all transition times be one unit long, we now let the amount of time spent in state i before making a transition to state j be a random variable with a general distribution function $F_{ij}(\bullet)$. With this change, we have converted our discrete-time Markov chain into a semi-Markov process. We can now investigate the limiting behavior of this semi-Markov process as time approaches infinity by analyzing its embedded Markov chain. Specifically, for a three condition class rangeland, if we denote the stationary probabilities of the semi-Markov process by P_i, $i = 1,2,3$, it can be shown, using the methods outlined by Ross (1997, pp. 379-381), that these stationary probabilities satisfy $P_i = \pi_i \mu_i / \sum_{j=1}^{j=3} \pi_j \mu_j$, $i = 1,2,3$, where π_i is the embedded Markov chain's stationary probability of being in state i and μ_i is the mean time the semi-Markov process (rangeland) spends in state i.

3. Conclusions

The conceptual approach presented above highlights the advantages of linking policy to the dynamic and the stochastic structure of a rangeland. Specifically, our approach is consistent with the state-and-transition model of range behavior proposed by Westoby *et al.* (1989). This approach explicitly accounts for the fundamental role that uncertainty plays in the temporal evolution of managed rangelands. This approach also makes transparent the role that policy can play, and, on occasion not play, in ensuring the sustainable use of rangelands. Finally, the optimization of policy objectives ($\overline{D}$ and $\overline{U}$) arising from this approach involves the simultaneous maintenance of ecological stability in the sense of persistence.

The state and transition model can be extended in a number of ways such as those outlined in section 2.4 above. However, given the paucity of quality empirical research on the management of dynamic and stochastic rangelands, we now point out a salient empirical extension of this model. Transition probabilities are an important component of all the Markov models discussed above. Consequently, it would be useful to follow the lead of Ethridge *et al.* (1985) and estimate the transition probabilities of a Markov model of a parcel of rangeland. Knowledge of transition probabilities will enable a researcher to determine the practical merits of stationary versus non-stationary approaches to the study of dynamic and stochastic rangelands.

REFERENCES

Batabyal, A.A. 2000. An Interdisciplinary Research Agenda for the Study of Ecological-Economic Systems in the American West, *Resources Policy*, 26, 69-75.

Batabyal, A.A., Biswas, B., and Godfrey, E.B. 2001. The Stocking Rate versus Time in Range Management: A Review of Recent Theoretical Developments, *Applied Economics Letters*, 8, 555-559.

Ethridge, D.E., Roy, S.K., and Myers, D.W. 1985. A Markov Chain Analysis of Structural Changes in the Texas High Plains Cotton Ginning Industry, *Southern Journal of Agricultural Economics*, 17, 11-20.

Holechek, J.L., Pieper, R.D., and Herbel, C.H. 2001. *Range Management*, 4th ed. Prentice Hall, Englewood Cliffs, NJ.

Isaacson, D.L., and Madsen, R.W. 1976. *Markov Chains*. Wiley, New York, NY.

Lambert, D.K., and Harris, T.R. 1990. Stochastic Dynamic Optimization and Rangeland Investment Decisions, *Western Journal of Agricultural Economics*, 15, 186-195.

McCluskey, J.J., and Rausser, G.C. 1999. Federal Grazing Reform and Avoidable Risk, *Journal of Agricultural and Resource Economics*, 24, 140-154.

Passmore, G., and Brown, C. 1991. Analysis of Rangeland Degradation Using Stochastic Dynamic Programming, *Australian Journal of Agricultural Economics*, 35, 131-157.

Perrings, C. 1998. Resilience in the Dynamics of Economy-Environment Systems, *Environmental and Resource Economics*, 11, 503-520.

Pimm, S.L. 1991. *The Balance of Nature?* University of Chicago Press, Chicago, IL.

Ross, S.M. 1997. *Introduction to Probability Models*, 6th ed. Academic Press, San Diego, CA.

Westoby, M., Walker, B., and Noy-Meir, I. 1989. Opportunistic Management for Rangelands Not at Equilibrium, *Journal of Range Management*, 42, 266-274.

Chapter 8

THE CHOICE BETWEEN THE STOCKING RATE AND TIME IN RANGE MANAGEMENT

with B. Biswas and E. B. Godfrey

ABSTRACT

A long standing question in range management concerns the relative importance of the *stocking rate* versus the length of *time* during which animals graze a particular rangeland. We address this question by analyzing the problem faced by a private rancher who wishes to minimize the long-run expected net unit cost ($LRENC$) from range operations by choosing either the stocking rate or the length of time during which his animals graze his rangeland. We construct a renewal-theoretic model and show that, in general, this rancher's $LRENC$ with an optimally chosen stocking rate is *lower* than his $LRENC$ with an optimally chosen grazing cycle length. From a management perspective, this means that correct stocking of the range is more important than the length of time during which animals graze the range. In addition, our research shows how to address questions concerning the desirability of temporal versus nontemporal controls in managing natural resources such as fisheries and hunting grounds.

Keywords: Long Run Expected Net Cost, Range Management, Stocking Rate, Time

1. INTRODUCTION

1.1. Preliminaries

All parts of the world that are not bare deserts, that are not cultivated, and that are not covered by bare soil, ice, or rock can be thought of as *rangelands*. This means that rangelands include most deserts, forests, and all natural grasslands. The key feature of a rangeland is that it consists of uncultivated land that can and typically does provide habitat for browsing and grazing animals. *Browsing* refers to the consumption of leaves and twigs from woody plants such as shrubs and trees by animals. In contrast, *grazing* refers to the consumption of standing forage such as grasses by animals.

Range management is "the manipulation of rangeland components to obtain the optimum combination of goods and services for society on a sustained basis" (Holechek *et al.* 1998, p. 5). As noted by Stoddard *et al.* (1975, pp. 2-3) and by Holechek *et al.* (1998, p. 5), in contemporary times, the task of range management is based on five basic precepts. First, a rangeland is a renewable resource. Second, solar energy captured by the green plants of a rangeland can only be harvested by browsing and grazing animals. Third, the productivity of a rangeland is determined by climatic, soil, topographic, and use factors. Fourth, in comparison with cultivated lands, rangelands provide humans with food and fiber at very low energy costs. Finally, a variety of goods and services such as food, minerals, timber, and recreation are obtained from rangelands.

A range manager can manipulate the components of a rangeland in several ways. In other words, this manager can accomplish his managerial objectives[1] with a variety of choice variables. In this chapter, we are interested in shedding light on a particular controversy in the range management literature. This controversy concerns two choice variables, namely, the stocking rate and the length of a grazing cycle. The *stocking rate* concept is used in more than one way by range managers. Consequently, it is important to be clear about the precise meaning of this concept. The meaning that we shall use in this chapter tells us that the "stocking rate is typically expressed as animal units per section of land" (Holecheck *et al.* 1998, p. 190). The *length of a grazing cycle* is more straightforward and it is defined to be the length of time in a calendar year during which animals graze a given rangeland.[2]

With these two definitions in place, we are now in a position to state the above mentioned controversy in the form of a simple question: Is the stocking rate more important or is time more important in range management? The objective of this chapter is to answer this question. We now discuss this range management controversy in greater detail and then we comment on the way in which we plan to address the underlying issues.

1.2. The Controversy

Although there are many aspects to the task of range management, today, range scientists agree that one important aspect concerns the determination of the appropriate stocking rate. Consider the position of two standard range management texts on the subject of the stocking rate. Stoddart *et al.* (1975, p. 262) tell us that correct livestock "numbers are important for the perpetuation of the range, the well-being of the livestock, and the economic stability of the operator." Holecheck *et al.* (1998, p. 221) go even further and state that proper "stocking is the most important part of successful range management."

However, not everyone agrees that the stocking rate is the most salient part of successful range management. In particular, Allan Savory (1983, 1988) and his adherents — see Goodloe (1969), Savory and Parsons (1980), and Savory and Butterfield (1999) — have

[1] Examples include the maximization of (i) range livestock productivity, and (ii) the economic returns from the rangeland.

[2] In the rest of this chapter, we shall use the terms "length of grazing cycle" and "time" interchangeably. The reader should note that both these terms refer to the length of time during which animals graze a given rangeland.

forcefully argued that the stocking rate is *less* important than is commonly believed. Savory and Butterfield (1999, p. 41, emphasis in original) have pointed out that until "very recently no one truly explored the question of *when* animals are there as opposed to how many there are." The central point of Allan Savory and other like minded scholars is this: Overgrazing bears "little relationship to the number of animals but rather to the *time* plants [are] exposed to the animals" (Savory and Butterfield 1999, p. 46, emphasis in original).[3]

This polarized state of affairs raises an important question. Is the stocking rate more important or is time, i.e., the length of the grazing cycle, more important in range management? We shall answer this question by analyzing the decision problem faced by an optimizing private rancher. Keeping with standard practice in economics, we suppose that this rancher wishes to maximize the long-run expected profit — or equivalently, minimize the long-run expected net unit cost (hereafter $LRENC$) — from his range operations. This rancher does so by choosing either the stocking rate or the length of time during which his animals graze his rangeland.[4] Note that this *long run* focus means that our rancher cares about the expected net cost from his range operations *and* about the well-being of his rangeland. Moreover, this focus also implies that the rancher will not stock his rangeland at a rate that the rangeland is unable to support. Our analysis shows that, in general, this rancher's $LRENC$ with an optimally chosen stocking rate is *lower* than his $LRENC$ with an optimally chosen grazing cycle length.

To intuitively see why this result holds, note the following two things. First, because the two choice variables under consideration here are different — the stocking rate is a *quantity* control variable and the grazing cycle length is a *temporal* control variable — the rancher's objective functions with these two choice variables are dissimilar. Second, in the presence of uncertainty, the two choice variables under consideration affect the rancher's objective function in different ways. The net impact of these two things is that our rancher's minimized $LRENC$ with an optimally chosen grazing cycle length is higher than his minimized $LRENC$ with an optimally chosen stocking rate by a specific additive factor. This additive factor is $c/2$ where c can be thought of as the instantaneous net cost of grazing an animal.

At the outset, it is important to be clear about the conclusions that can be drawn from our analysis. Our analysis tells us that when confronted with a choice *between* the stocking rate and the grazing cycle length, a rational rancher would choose the stocking rate. We are *not* saying that the grazing cycle length is irrelevant for management purposes. Further, it is our conjecture that just as part-price and part-quantity control instruments dominate pure price and pure quantity control instruments (see Roberts and Spence 1976; Weitzman 1978; Batabyal 1995), a hybrid control instrument that is part-stocking rate and part-time is likely to be more useful for range management than the stocking rate or the grazing cycle length alone. Having said this, we should note that the simultaneous use of the two control variables under consideration here is not mandatory. Just as it is not always necessary to use price and

---

[3] Allan Savory's views on grazing have been variously described as time-controlled grazing, as short-duration grazing, and as the Savory grazing method. For more on this and related issues, see Holecheck *et al.* (1998, pp. 229-256). A related issue here concerns plant recovery times after different degrees of grazing. For more on this, see Hart *et al.* (1988), Hall *et al.* (1992), and McCreary and Tecklin (1993).

[4] To keep the mathematics straightforward, in the rest of this chapter, we shall not focus on the profit criterion; instead, we shall focus on the $LRENC$ criterion. However, the reader should note that maximizing profit is equivalent to minimizing net cost.

quantity control instruments simultaneously to regulate pollution, similarly, in this range management context, it is not imperative that a range manager use a temporal control (grazing cycle length) and a quantity control (stocking rate) concurrently.

Given the obvious importance of this stocking rate versus time question for practical range management, one would expect this question to have been studied thoroughly. Although there are many studies that have evaluated the impact of alternate stocking rates on animal performance and on forage production,[5] and some empirical studies of Allan Savory's time-controlled grazing,[6] these studies have not resolved this stocking rate versus time controversy. Moreover, on the theoretical side, the matter is even less settled. To the best of our knowledge, there are *no* previous theoretical studies of this question. This state of affairs has led Holechek *et al.* (1998, p. 254) to conclude that the long "term impacts of [time-controlled] grazing . . . [have yet] to be determined." As such, we now proceed to our analysis of the long run effects of the stocking rate versus time in range management.

The theoretical framework of this chapter is adapted from Batabyal (1999) and the rest of this chapter is organized as follows. Section 2 presents and analyzes a renewal-theoretic[7] model of the decision problem faced by a private rancher who wishes to minimize his $LRENC$ from range operations by choosing the stocking rate optimally. Section 3 analyzes a similar model; however, in this section, the rancher minimizes his $LRENC$ from range operations by choosing the length of the grazing cycle optimally. Section 4 first compares the optimized value of the rancher's $LRENC$ from sections 2 and 3 and thereby determines which choice variable — stocking rate or time — results in lower $LRENC$. Next, this section discusses the relationship between the analysis of this chapter and other related natural resource management problems. Section 5 concludes and offers suggestions for future research.

2. Range Management with an Optimally Chosen Stocking Rate

Consider a private rancher who owns livestock animals (cows) and a fenced plot of rangeland. In the model of this and the next section, our private rancher conducts his range operations with reference to a particular grazing period in a calendar year. For instance, this grazing period might be from May 2 to July 15, which would correspond to the grazing period for intensive-early stocking, or it might run from May 2 to October 3, which would correspond to the grazing period for normal season-long grazing (Holechek *et al.* 1998, pp. 231-236). At the beginning of a grazing period, our rancher lets his animals into his fenced rangeland in accordance with an arrival process. In general, this arrival process could be any renewal process. However, in both the economics and the ecology literatures, the Poisson process has been frequently used to study natural resource phenomena.[8] Consequently, we

---

5 See Holechek *et al.* (1998, pp. 248-256) and Holecheck *et al.* (1999).

6 See Graham *et al.* (1992), Hart *et al.* (1993), and Holechek *et al.* (1998, pp. 248-256).

7 For more on renewal theory, see Ross (1996, pp. 98-161; 1997, pp. 351-410) and Taylor and Karlin (1998, pp. 419-472).

8 See Uhler and Bradley (1970), Pielou (1977), Arrow and Chang (1980), Mangel (1985), and Batabyal and Beladi (2001) for a more detailed corroboration of this claim.

suppose that this arrival process is the Poisson process with rate α.[9] This rancher believes that the appropriate stocking rate for his rangeland corresponds to A animals. As such, once A animals have been allowed into the rangeland to graze, entry of additional animals is prohibited for the grazing period under consideration. Put differently, once A animals have been allowed in, this rancher's rangeland is closed to grazing in the current calendar year grazing period.

As a result of his range operations, our rancher incurs costs and obtains benefits from two sources. The first, or direct, source of net cost (total cost less total benefit) stems from things like the deleterious effects of grazing on the plant species of the rangeland (a cost) and from the weight gain accruing to animals as a result of forage intake (a benefit). We capture this direct source of net cost by supposing that our rancher incurs net cost at the rate \$$ac$ per unit time, where a refers to the number of animals grazing at that time and c can be thought of as the instantaneous net cost per animal. The second, or indirect, source of net costs arises from things like the need to feed animals that have not been allowed into the rangeland (a cost) and from stocking the rangeland at the correct rate (a benefit). This benefit arises because correct stocking means that the rangeland's grazing capacity will not be exceeded. In turn, this means that this rangeland will be able to provide the rancher's animals with a flow of forage in the long run. In every calendar year grazing period, we suppose that our rancher incurs a net cost of \$$C$, when he closes his rangeland to additional animals.

Now, if we say that a grazing cycle[10] for the calendar year is completed whenever the rancher closes the rangeland to additional animals, then the description of events in the previous two paragraphs constitutes a renewal-reward process.[11] This fact is useful because we can now use a key property of renewal-reward processes, namely, the renewal-reward theorem to compute our rancher's $LRENC$ from his range operations. The reader should note that the rancher's objective function involves the minimization of an economic criterion, i.e., the long-run expected net cost. The renewal-reward theorem tells us how to compute this economic criterion. Specifically, this theorem tells us that the rancher's $LRENC$ equals the expected net cost in a grazing cycle divided by the expected length of this grazing cycle. Formally, we have

---

[9] For more on the Poisson process, see Ross (1996, pp. 59-97; 1997, pp. 249-301) and Taylor and Karlin (1998, pp. 267-332). The rate of the (Poisson) arrival process might depend on the stock of animals. One way to model this would be to work with a nonhomogeneous Poisson process for which the arrival rate at time t is a function of t. That is, the arrival rate (also called the intensity function) is $\alpha(t), t \geq 0$. Now if this intensity function is bounded, i.e., if $\alpha(t) \leq \beta\ \forall t$, then we can think of the nonhomogeneous Poisson process as being a random sample from the homogeneous Poisson process with rate β. Specifically, we could work with this new Poisson process and the analysis would go through as indicated in the chapter. Finally, note that the range management problem being analyzed in this chapter is directly concerned with the number of animals and the length of time during which animals are *on the rancher's rangeland*. The question of how the animals leave the rangeland is not of interest. Formally, the problem being analyzed is *not* a queuing problem. As such, it is not necessary to formally model the departure process.

[10] The use of the word "cycle" is appropriate because the events that we are studying here are cyclical in nature. First, with regard to the analysis in this section, a cycle is completed whenever the rangeland is closed to grazing by animals. Similarly, with respect to the analysis in section 3, a cycle is completed whenever T for that calender year expires. Second, this pattern of events is repeated every calender year and we are examining the long-run behavior of this cyclical pattern of events.

[11] For more on renewal-reward processes and the renewal-reward theorem, see Ross (1996; 1997) and Taylor and Karlin (1998). Also, see the motivation for the rancher's objective function in section 1.2.

$$LRENC = \frac{E[net\ \cos t\ per\ grazing\ cycle]}{E[length\ of\ \ grazing\ cycle]} \tag{1}$$

where $E[\bullet]$ is the expectation operator.

Let us now compute the two expectations on the right-hand-side (hereafter RHS) of equation (1). In any given grazing cycle, let X_a denote the time between the arrival of the ath animal and the $(a+1)^{\text{th}}$ animal into the rancher's rangeland. Then the numerator on the RHS of equation (1) is given by

$$E[net\ \cos t\ per\ grazing\ cycle] = C + E[1cX_1 + 2cX_2 + 3cX_3 + ... + (A-1)cX_{A-1}]. \tag{2}$$

Because the rancher's cows are brought into the rangeland in accordance with a Poisson process with rate α, the mean interarrival time is $1/\alpha$. Mathematically, this means that $E[X_i] = 1/\alpha, i = 1, ..., (A-1)$. Using this result, the RHS of equation (2) can be simplified to

$$E[net\ \cos t\ per\ grazing\ cycle] = C + \frac{cA(A-1)}{2\alpha}. \tag{3}$$

In order to compute the denominator on the RHS of equation (1), it suffices to note that the expected length of a grazing cycle is simply the expected time it takes for the animals to begin grazing on the rancher's rangeland. Because the mean interarrival time for the cows is $1/\alpha$ we get

$$E[length\ of\ \ grazing\ cycle] = \frac{A}{\alpha}. \tag{4}$$

Now, combining the results from equations (3) and (4), we get an expression for the rancher's $LRENC$. That expression is

$$LRENC = \frac{\alpha C}{A} + \frac{c(A-1)}{2}. \tag{5}$$

Having computed the expression for our rancher's $LRENC$, we are now in a position to state this rancher's $LRENC$ minimization problem. Specifically, this rancher chooses the stocking rate A to minimize the $LRENC$ from his range operations. Formally, our rancher solves

$$\min_{(A)} \left[\frac{\alpha C}{A} + \frac{c(A-1)}{2} \right]. \tag{6}$$

Treating A as a continuous choice variable and using calculus, we see that the stocking rate that minimizes the rancher's $LRENC$ is given by[12]

$$A^* = \sqrt{\frac{2\alpha C}{c}}. \tag{7}$$

In other words, the optimal stocking rate equals the square root of the ratio of the product of twice the rate of the Poisson arrival process (α) and the indirect net cost from closing the rangeland to additional animals (C) to the instantaneous net cost per animal (c). Inspecting equation (7), it is easy to verify two properties of the optimal stocking rate. First, as the indirect net cost per grazing cycle (C) goes up, the rancher finds it desirable to *raise* the optimal stocking rate. Second, if the instantaneous net cost per animal (c) increases, then it is in the interest of the rancher to *lower* the optimal stocking rate.

Let us now substitute the expression for the optimal stocking rate from equation (7) into the minimand in equation (6). This gives us an expression for the minimal $LRENC$ that our rancher will incur by choosing the stocking rate optimally. Denote this minimal $LRENC$ by $(LRENC)^*_{SR}$. Some algebra tells us that

$$(LRENC)^*_{SR} = \sqrt{2\alpha c C} - \frac{c}{2}. \tag{8}$$

Inspecting equation (8), we see that the minimal $LRENC$ that our rancher will incur by choosing the stocking rate optimally equals the square root of the product of twice the rate of the Poisson arrival process (α), the instantaneous net cost per animal (c), and the indirect net cost per grazing cycle (C), less one-half the instantaneous net cost per animal.

We now study the case in which the focus of our private rancher is not on the stocking rate *per se*, but on the *length* of the grazing cycle on his rangeland. After computing the optimal length of the grazing cycle, we shall compare equation (8) with the corresponding equation for this latter case in which the rancher's focus is on time.

3. Range Management with an Optimally Chosen Grazing Cycle Length

Instead of choosing the stocking rate optimally, our rancher now follows a different strategy. In particular, this rancher now chooses the length of the grazing cycle (T) to minimize the $LRENC$ from his range operations. In the context of the discussion in the first paragraph of section 2, this means that if the grazing period in a calendar year happens to be 75 days long (May 2 to July 15), then our rancher chooses T with this 75-day grazing period in mind.[13] So, in this example, the optimal T would be some real number between 0 and 75. If

[12] The second-order condition is satisfied.

[13] The reader should note that the length of the grazing period in a calender year and the choice of T will depend, *inter alia*, on the geographic location of the rangeland in question. Specifically, the recovery time

the optimal $T = 0$, then this means that the rancher rests his rangeland for the entire grazing period in that calendar year. At the other end, if optimal $T = 75$, then this means that the rancher's grazing cycle and the grazing period for that calendar year coincide.

In this setting, our rancher chooses the length of the grazing cycle (time) to minimize the $LRENC$ from his range operations. Consequently, let us now compute the $LRENC$ that is incurred by the rancher when this rancher's focus is on time rather than on the stocking rate. As in the previous section, at the beginning of the grazing period, our rancher lets his animals into his rangeland in accordance with a Poisson process with rate α. We suppose that this rancher lets his animals graze the rangeland for T units of time. In other words, when T units of time have elapsed, the rangeland is closed to grazing. This means that a grazing cycle is completed when T units of time have elapsed. As explained in the previous paragraph, the length of this grazing cycle will either be less than or equal to the length of the grazing period in a calendar year.

We shall use the renewal-reward theorem (equation (1)) to compute our rancher's $LRENC$. The computation of $E[net\ cost\ per\ grazing\ cycle]$ will be facilitated by conditioning on $N(T)$, the total number of animals that are grazing the rancher's rangeland by time T. This yields

$$E[net\ cost\ per\ grazing\ cycle / N(T)] = C + \frac{cTN(T)}{2}. \tag{9}$$

Using the properties of the expectation operator and equation (9), we get

$$E[net\ cost\ per\ grazing\ cycle] = C + \frac{\alpha c T^2}{2}. \tag{10}$$

Now note that $E[net\ cost\ per\ grazing\ cycle] = T$. This result and equation (10) together tell us that our rancher's $LRENC$ is given by

$$LRENC = \frac{C}{T} + \frac{\alpha c T}{2}. \tag{11}$$

Having computed the expression for our rancher's $LRENC$, we are now in a position to state this rancher's $LRENC$ minimization problem. This rancher chooses the length of the grazing cycle (T) to minimize the $LRENC$ from his range operations. Formally, our rancher solves[14]

---

and the capability of a range in a wet and humid region will be very different from the recovery time and capability of a range in an arid and/or semiarid region.

[14] To keep this minimization problem simple, we have not imposed a constraint requiring T to be bounded below by zero and above by the length of the grazing period in a calender year. If the optimal T turns out to be larger than the length of the grazing period, then we simply set the optimal T equal to the length of the grazing period. For example, as discussed in the first paragraph of this section, if the length of the grazing period happens to be 75 days and the optimal T turns out to be 78 days, then we simply set this optimal T equal to 75 days.

$$\min_{\{T\}}\left[\frac{C}{T}+\frac{\alpha cT}{2}\right]. \tag{12}$$

Using calculus, we see that the grazing cycle length that minimizes the ranchers's *LRENC* is given by[15]

$$T^{*}=\sqrt{\frac{2C}{\alpha c}}. \tag{13}$$

In other words, the optimal length of the grazing cycle equals the square root of the ratio of the product of twice the indirect net cost from closing the rangeland (C) to the product of the rate of the Poisson arrival process (α) and the instantaneous net cost per animal (c). Inspecting equation (13) it is easy to verify two properties of the optimal length of the grazing cycle. First, as the indirect net cost per grazing cycle (C) goes up, the rancher finds it optimal to *lengthen* the grazing cycle. Second, if the instantaneous net cost per animal (c) increases, then it is optimal for the rancher to *shorten* the grazing cycle.

Let us now substitute the expression for the optimal length of the grazing cycle from equation (13) into the minimand in equation (12). This gives us an expression for the minimal *LRENC* that our rancher will incur by choosing the grazing cycle length optimally. Denote this minimal *LRENC* by $(LRENC)_{T}^{*}$. After some algebra, we get

$$(LRENC)_{T}^{*}=\sqrt{2\alpha cC}. \tag{14}$$

Inspecting equation (14), we see that the minimal *LRENC* that our rancher will incur by choosing the grazing cycle length optimally equals the square root of the product of twice the rate of the Poisson arrival process (α), the instantaneous net cost per animal (c), and the indirect net cost per grazing cycle (C).

Recall that the objective of this chapter is to answer the following question: Is the stocking rate more important or is time, i.e., the length of the grazing cycle, more important in range management? We now provide an answer to this question.

4. Stocking Rate versus Time in Range Management

Equation (8) gives us an expression for the *LRENC* incurred by our rancher when he chooses the stocking rate optimally. Similarly, equation (14) gives us an expression for this rancher's *LRENC* when he chooses the length of the grazing cycle (time) optimally. Comparing these two expressions, we see that

$$(LRENC)_{SR}^{*}=\sqrt{2\alpha cC}-\frac{c}{2}<\sqrt{2\alpha cC}=(LRENC)_{T}^{*}. \tag{15}$$

---

[15] The second-order condition is satisfied.

Equation (15) clearly tells us that the rancher's *LRENC* with an optimally chosen stocking rate is lower than his *LRENC* with an optimally chosen grazing cycle length. It is in this sense that the stocking rate is *more important* than time in range management. Put differently, if a rational rancher had to choose a single control variable from a control set consisting of the stocking rate and time, then this rancher would choose the stocking rate over time. We now discuss the relationship between the analysis of this chapter and other related natural resource management problems.

4.1. Our Analysis and Other Resource Management Problems

In addition to rangelands, a number of other natural resources are also managed with temporal and nontemporal choice variables. For instance, commercial and recreational hunters for most game are subject to seasonal (time) restrictions. Moreover, such hunters are generally required to hunt during daylight hours. Similarly, Batabyal and Beladi (2002) have pointed out that most commercial fisheries are subject to season length (time) restrictions. Given this state of affairs, it would certainly be useful to know whether society is better off with such temporal restrictions or whether nontemporal choice variables — such as the number of animals hunted and the number of fishing boats used — result in higher welfare to society.

The theoretical framework of this chapter can be used to answer these sorts of questions. Specifically, in the context of range management decisions, our analysis leads to three conclusions. First, *ceteris paribus*, correct stocking of the range is more important than the length of time during which animals graze the range. This conclusion supports the view that proper "stocking is the most important part of successful range management" (Holechek *et al.* 1998, p. 221). Second, there are circumstances in which the use of a nontemporal choice variable like the stocking rate leads to lower costs for the rancher. As such, it would be useful to see if one can make a general theoretical argument against the use of temporal choice variables in range management. Finally, although we have come down on the side of the stocking rate, it is clear that because of biological factors such as the differential recovery rates of plants subject to grazing, there is a difference between grazing 10 animals for 100 days and grazing 100 animals for 10 days. In other words, the length of the grazing cycle is a relevant choice variable. This suggests that from a management perspective, stochastic but relevant biological factors are likely to be better accounted for by the use of control instruments that are part-stocking rate and part-time. We now discuss this issue in greater detail.

5. Conclusions

In this chapter, we used a renewal-theoretic approach to analyze the decision problem faced by a private rancher who is interested in minimizing the *LRENC* from his range operations. On the basis of our analysis of two optimization problems for this rancher, we concluded that the stocking rate is more important than time in range management. To the best of our knowledge, this is the *first* theoretical answer to this stocking rate versus time question in range management.

The analysis of this chapter can be extended in a number of directions. In what follows, we suggest two possible extensions. First, the discussion in the last paragraph of Section 4.1 suggests that there might exist choice variables, intermediate between the stocking rate and time, that dominate these two control variables. With regard to this issue, consider the seminal work of Roberts and Spence (1976). Environmental economists now know that is possible to construct an "intermediate" control instrument that is part-price (fee or tax) and part-quantity (emissions permit scheme). Roberts and Spence (1976) showed that this intermediate control instrument can always be converted into a pure price or pure quantity control instrument. Consequently, in comparison with either a pure price or pure quantity control instrument, a regulator will do at least as well — and often much better — with this intermediate control instrument. A useful extension of this chapter would be to determine whether this logic carries over to the subject of range management. In other words, the open question is to check whether it is possible to construct, in a dynamic and stochastic setting, a control instrument that is intermediate in the sense that it is part stocking rate and part time. If it is possible to do so, then it should be fairly straightforward to demonstrate that this intermediate control instrument dominates a pure stocking rate and a pure time control instrument.

Second, in section 2, we studied the decision problem faced by a private rancher who owns a single species of livestock animals (cows). As such, it would be useful to ascertain whether the results of this chapter hold when this rancher's decision problem with the stocking rate as a choice variable is modified to account for situations in which the rancher owns more than one animal species. Studies of range management that incorporate these aspects of the problem into the analysis will provide additional insight into the roles that the stocking rate and time play in successful range management.

REFERENCES

Arrow, K.J., and Chang, S.S.L. 1980. Optimal Pricing, Use, and Exploration of Uncertain Natural Resource Stocks, in P.T. Liu (Ed.), *Dynamic Optimization and Mathematical Economics*. Plenum Press, New York, NY.

Batabyal, A.A. 1995. Leading Issues in Domestic Environmental Regulation: A Review Essay, *Ecological Economics*, 12, 23-39.

_____ . 1999. Aspects of the Optimal Management of Cyclical Ecological-Economic Systems, *Ecological Economics*, 30, 285-292.

Batabyal, A.A., and Beladi, H. 2001. Aspects of the Theory of Financial Risk Management for Natural Disasters, *Applied Mathematics Letters*, 14, 875-880.

_____ . 2002. Time Restrictions in Natural Resource Management: A Dynamic and Stochastic Analysis. Forthcoming, *European Journal of Operational Research*.

Goodloe, S. 1969. Short Duration Grazing in Rhodesia, *Journal of Range Management*, 22, 369-373.

Graham, K.T., Torell, L.A., and Allison, C.D. 1992. Costs and Benefits of Implementing Holistic Resource Management on New Mexico Ranches, *New Mexico Agricultural Experiment Station Bulletin 672*.

Hall, L.M., George, M.R., McCreary, D.D., and Adams, T.E. 1992. Effects of Cattle Grazing on Blue Oak Seedling Damage and Survival, *Journal of Range Management*, 45, 503-506.

Hart, R.H., Samuel, M.J., Test, P.S., and Smith, M.A. 1988. Cattle Vegetation and Economic Responses to Grazing Systems and Grazing Pressure, *Journal of Range Management*, 41, 282-286.

Hart, R.H., Clapp, S., and Test, P.S. 1993. Grazing Strategies, Stocking Rates and Frequency Intensity of Grazing on Western Wheatgrass and Blue Grama, *Journal of Range Management*, 46, 122-127.

Holechek, J.L., Pieper, R.D., and Herbel, C.H. 1998. *Range Management*, 3rd edition. Prentice Hall, Upper Saddle River, NJ.

Holechek, J.L., Thomas, M., Molinar, F., and Galt, D. 1999. Stocking Desert Rangelands: What We've Learned, *Rangelands*, 21, 8-12.

Mangel, M. 1985. *Decision and Control in Uncertain Resource Systems*. Academic Press, Orlando, FL.

McCreary, D.D., and Tecklin, J. 1993. Tree Shells Accelerate Valley Oak Restoration on Grazed Rangelands (California), *Restoration and Management Notes*, 11, 152.

Pielou, E.C. 1977. *Mathematical Ecology*. Wiley, New York City, NY.

Roberts, M., and Spence, M. 1976. Effluent Charges and Licenses Under Uncertainty, *Journal of Public Economics*, 5, 193-208.

_____ . 1996. *Stochastic Processes*, 2nd edition. Wiley, New York City, NY.

Ross, S.M. 1997. *Introduction to Probability Models*, 6th edition. Academic Press, San Diego, CA.

Savory, A. 1983. The Savory Grazing Method or Holistic Resource Management, *Rangelands*, 5, 155-159.

_____ . 1988. *Holistic Resource Management*. Island Press, Washington, DC.

Savory, A., and Parsons, S.D. 1980. The Savory Grazing Method, *Rangelands*, 2, 234-237.

Savory, A., and Butterfield, J. 1999. *Holistic Management*, 2nd edition. Island Press, Washington, DC.

Stoddart, L.A., Smith, A.D., and Box, T.W. 1975. *Range Management*, 3rd edition. McGraw Hill, New York, NY.

Taylor, H.M., and Karlin, S. 1998. *An Introduction to Stochastic Modeling*, 3rd edition. Academic Press, San Diego, CA.

Uhler, R.S., and Bradley, P.G. 1970. A Stochastic Model for Determining the Economic Prospects of Petroleum Exploration Over Large Regions, *Journal of the American Statistical Association*, 65, 623-630.

Weitzman, M.L. 1978. Optimal Rewards for Economic Regulation, *American Economic Review*, 68, 683-691.

Chapter 9

AN UPPER BOUND FOR THE TRANSIENT RESILIENCE OF A RANGELAND THAT IS MANAGED USING CELL GRAZING

ABSTRACT

Ecologists and economists have noted that the stability property known as resilience has an asymptotic and a transient interpretation. Further, several recent studies have analyzed the asymptotic and the transient resilience of jointly determined ecological-economic systems. This notwithstanding, to the best of our knowledge, there are no theoretical studies of either the asymptotic or the transient resilience of rangelands that are managed using cell grazing. Therefore, this chapter has two objectives. First, we use a stochastic framework to model a rangeland that is managed using cell grazing. Next, we compute an upper bound for the transient resilience of this rangeland.

Keywords: Cell Grazing, Rangeland, Stochastic, Transient Resilience

1. INTRODUCTION

Ecologists and economists now agree that rangelands are jointly determined ecological-economic systems and that they ought to be studied as one system. Given this recognition, a number of questions about the management of rangelands become salient. Specifically, it is important to know how rangelands respond to shocks. In the ecology literature, the response of ecosystems to shocks has often been measured by the notion of resilience. As discussed in Neubert and Caswell (1997), resilience is measured by first constructing a matrix that gives the strength of the interactions between the various species at their equilibrium population densities. The largest eigenvalue of this matrix is then defined to be the resilience of this ecosystem. In other words, this notion of resilience measures the speed with which a stable ecosystem returns to its original state following a shock.[1]

---

[1] It is important to point out that two notions of resilience are prevalent in the ecology literature. The notion that we have just mentioned is resilience in the sense of Pimm (1984). In contrast, resilience in the sense of Holling refers to "the amount of disturbance that can be sustained [by an ecosystem] before a change in

Although this method for calculating resilience has been frequently used in the ecology literature, it is important to note that resilience, measured in terms of the largest eigenvalue of the matrix of interaction effects, is an *asymptotic* property. In other words, this notion of resilience provides information about the rate of decay of shocks as time approaches infinity. Moreover, this notion of resilience completely ignores the short-term, transient behavior of an ecosystem such as a rangeland. However, for rangelands in particular and ecological-economic systems in general, it is transient behavior that is of interest. As Batabyal (1998, 1999, 2000) and others have noted, knowing how a given rangeland responds to shocks in the long run, i.e., as time approaches infinity, would not be particularly useful because all range managers set policy over much shorter time horizons. In addition to this, Neubert and Caswell (1997) have convincingly argued that ecosystems cannot be managed effectively without understanding their transient behavior. Specifically, these researchers have shown that "even in stable, resilient [ecosystems], transient behavior can be dramatic, long lasting, and counterintuitive" (Neubert and Caswell, 1997, p. 654).

Given the salience of *transient* behavior for the effective management of ecological-economic systems, one can ask what is known about transient resilience in the context of rangelands. Unfortunately, the answer is very little. As noted in Batabyal (2000), the economics literature appears to have ignored this issue completely. Even in the range ecology literature, there are very few studies of rangeland resilience. Bork *et al.* (1998) and Perevolotsky *et al.* (2001) have studied, respectively, the resilience of a three-tip sagebrush range in Idaho and the resilience of the prickly burnet in eastern Mediterranean ranges. However, both these studies are empirical in nature and they do not employ an explicitly stochastic framework. Moreover, neither of these two studies are concerned with cell grazing.[2]

Given this state of affairs, the present chapter has two objectives. First, we use a stochastic framework to model a rangeland that is managed using cell grazing. Next, we compute an upper bound for the transient resilience of this rangeland. We focus on cell grazing because it is commonly used in the United States and in other parts of the world (see Holechek *et al.*, 2001, p. 249) and also because many of the characteristics of this kind of grazing are to be found in other grazing systems as well.[3] For the kind of model we use in this

system control or structure occurs" (Holling *et al.*, 1995, p. 50). The Holling notion of resilience was first proposed in Holling (1973). The Pimm (1984) notion of resilience focuses on stability near an equilibrium stationary state. Resistance to shocks and speed of return to the equilibrium are used to measure this notion of resilience. In contrast, the Holling (1973) notion of resilience focuses on conditions far away from an equilibrium stationary state where shocks can flip an ecosystem into another stability domain. In a recent contribution, Holling (1996) has referred to the Pimm (1984) notion of resilience as "engineering resilience" and to his own 1973 notion of resilience as "ecological resilience."

2 There is a sizeable literature in general ecology on resilience. However, most of this literature seems to have focused on asymptotic and on deterministic measures of resilience (see Ives (1995) and Neubert and Caswell (1997)). As a result, this body of research is unable to address the criticism that resilience measures need "to be defined for stochastic systems in which environmental perturbations are continuous..." (Ives, 1995, p. 217).

3 Cell grazing is also known as short duration grazing, as rapid rotation grazing and as time controlled grazing. For more on this and related matters see Savory and Butterfield (1999), Holechek *et al.* (2001, pp. 269-277), Batabyal (2001, also chapter 3) and Batabyal and Beladi (2002). The reader should note that other grazing systems such as the Merrill three-herd and four-pasture system and rest-rotation grazing also have rotational aspects to them.

chapter, it is not possible to obtain a closed-form expression for a rangeland's transient resilience. This is why we focus on an upper bound for a rangeland's transient resilience.

The rest of this chapter is organized as follows: Section 2 first describes a stochastic framework for analysis, it then uses this framework to compute the upper bound, and finally it discusses ways of operationalizing this upper bound. Section 3 concludes and offers suggestions for future research on the computation of the transient resilience of rangelands that are managed using cell grazing.

2. COMPUTING THE UPPER BOUND

2.1. Preliminaries

The basic idea in cell grazing is threefold. First, the manager divides the rangeland (or cell)[4] under study into a number of fenced paddocks of equal grazing capacity that are arranged like a wagon-wheel with water and animal handling facilities located at the center of the cell (Holechek *et al.*, 2001, p. 269). Next, the manager allows his or her herd of animals to graze a particular paddock for a specific time period. Finally, the herd of animals is moved to the next paddock, and the manager continues this process in a circular manner.

Our task now is to model this description of cell grazing. To this end, consider a grazing cell that has been divided into $n \in 1$ paddocks. To account for the wagon-wheel like arrangement of the paddocks, suppose that these paddocks are arranged around a notional circle. Now, the success of cell grazing depends in large part on choosing the temporal control optimally. In other words, the manager has to ensure that his or her animals do not spend too much time on a specific paddock before they are moved to the next paddock. If the animals graze a particular paddock for too long then this paddock will be degraded and it may not recover in time for grazing in subsequent grazing seasons. Further, if too many paddocks are degraded in a grazing season then the grazing cell itself is likely to be useless for grazing in subsequent grazing seasons.

During the course of a grazing season, two kinds of shocks affect the paddocks of our cell. The first kind of shock stems from grazing. The second kind of shock is environmental in nature and it stems from things like droughts, fires, and pests. Due to the combination of grazing and environmental shocks, each paddock of our cell will independently be degraded with probability p. Now recall that the Holling (1973) notion of resilience refers to the amount of disturbance that can be sustained by a rangeland before a change in system control or structure occurs. So, the key question now is this: When will there be a change in system control or structure for our cell that is subject to grazing and environmental shocks?

Some reflection tells us that this is most likely to happen when a sufficient number of paddocks are degraded so that, for all practical purposes, the cell itself is degraded and hence it cannot be used for grazing in the next grazing season. The reader will note that in general we do not know what constitutes a sufficient number of degraded paddocks but if we denote the sufficient number of degraded paddocks by $k \in 1$ □ then it is clear that $k \leq n$. Consequently, we shall say that our cell is degraded if and only if k consecutive paddocks

[4] In the rest of this chapter we shall use the terms "cell" and "rangeland" interchangeably.

are degraded. Put differently, there will be a change in system control or structure for our cell when this cell is degraded and this happens when k consecutive paddocks are degraded.

Now, following the lead of Krebs (1985, p. 587), Perrings (1998), and Batabyal (2000), we shall think of the transient resilience of our cell as the probability that this cell will *not* be degraded after a grazing season. Mathematically, we have *Transient Resilience* ≡ *Prob{no degradation}*. We now compute an upper bound for this probability.

2.2. The Stochastic Framework

To compute the upper bound of interest, we shall first calculate a lower bound for *Prob{degradation}* and then use the fact that *Prob{no degradation}* = *1- Prob{degradation}*. To this end, number the n paddocks that are arranged around a notional circle and let X_i, $i=1,...,n$, be the indicator variable[5] for the event that paddock i is not degraded but its nearest k neighbors in the clockwise direction are all degraded. Further, let X_{n+1} be the indicator variable for the event that all the n paddocks are degraded. Then from the description of cell grazing given in the first three paragraphs of section 2.1, it follows that

$$E[X_i]=(1-p)p^k\,,\ 1\le i\le n\,,\text{ and } E[X_{n+1}]=p^n\,, \tag{1}$$

where $E[\bullet]$ is the expectation operator.

Let $N=\sum_{i=1}^{i=n+1}X_i$. In words, N is the sum of all the events that imply that our cell is degraded. We now need to determine the expected value of N conditional on the events $X_i=1$ and $X_{n+1}=1$. Using standard procedures for computing a conditional expectation,[6] we get

$$E[N/X_i=1]=1+(n-2k-1)^+(1-p)p^k\,,\ i=1,...,n \text{ and } E[N/X_{n+1}=1]=1\,, \tag{2}$$

where the superscript + on the term $(n-2k-1)$ means that we are only interested in the positive part of the term within the parentheses. Now using Proposition 3.2.3 (also known as the conditional expectation bound) in Ross (2002, p. 82) we can compute a tight[7] lower bound for *Prob{degradation}*, the probability that our rangeland that is managed using cell grazing is not resilient. We get

$$\text{Prob}\{\text{degradation}\}=\text{Prob}\{\text{N}>0\}\ge \text{p}^\text{n}+\frac{n(1-p)p^k}{1+(n-2k-1)^+(1-p)p^k}\,. \tag{3}$$

---

[5] For more on indicator variables see Ross (2002, pp. 5-6).

[6] See Ross (2002, pp. 20-32).

[7] It is possible to use bounds other than the "conditional expectation bound" to compute the required lower bound but these bounds are not very tight. For a more detailed discussion about the meaning of a tight bound and why the "conditional expectation bound" is tight, see Ross (2002, pp. 85-87).

We are now in a position to state the upper bound for our rangeland's transient resilience. Recall that this transient resilience is given by *Prob{no degradation}*. Now using the fact that *Transient Resilience* ≡ *Prob{no degradation}* = 1- *Prob{degradation}* in equation (3), we deduce that

$$1-p^n-\frac{n(1-p)p^k}{1+(n-2k-1)^+(1-p)p^k}\geq \textit{Transient Resilience.} \tag{4}$$

2.3. Discussion

Equation (4) gives us the upper bound for the transient resilience of a rangeland that is managed using cell grazing. We see that this upper bound is a function of three things. These three things are the total number of paddocks in the cell (n), the number of consecutive degraded paddocks that imply degradation of the entire cell (k), and the individual paddock degradation probability (p).

In order to compute the upper bound in equation (4), it will be necessary to obtain information on the triple (n,k,p). The total number of paddocks (n) can be determined on the basis of past managerial experience with the rangeland under study. Following the discussion in Holechek *et al.* (2001, pp. 269-271) and Batabyal (2001, also chapter 3), we expect that (n) will typically be some number in the 8-30 range.

The value of (k) too will generally be determined on the basis of actual experience with a particular rangeland. The thing to note here is that because of the varying ecological properties of alternate rangelands, the value of (k) will generally vary from one type of rangeland to another. Put differently, the values of (k) for tallgrass prairie (found in the central United States) and for California annual grassland (found west of the Sierra Nevada mountains in the United States) will typically not be identical.

The paddock degradation probability (p) will most likely be the hardest to quantify. However, because (p) depends on grazing and environmental shocks, it should be possible to use climatological, forage nutrition, and grazing data with extant binary choice models to estimate (p) econometrically. Once this is done, we will have operationalized our upper bound on the transient resilience of a rangeland that is managed using cell grazing.

3. Conclusions

In this chapter we noted that there are *no* theoretical studies of the transient resilience of rangelands that are managed using cell grazing. Therefore, we used a stochastic framework to model a rangeland that is managed using cell grazing. Next, we used this stochastic framework to compute an upper bound for the transient resilience of this rangeland.

The analysis of this chapter can be extended in a number of different directions. In what follows, we suggest one possible extension. We used a fixed probability (p) to model the likelihood of paddock degradation. Although this makes sense because the paddocks in cell grazing are generally of equal grazing capacity (see Holechek *et al.*, 2001, p. 269), it would be useful to determine how the results of this chapter change when the probability (p) is paddock specific so that in place of a fixed (p) we have n different probabilities p_i, $i = 1,...,n$. Studies that incorporate this aspect of the problem into the analysis will provide richer and possibly more realistic characterizations of the transient behavior of rangelands that are managed using cell grazing.

REFERENCES

Batabyal, A.A. 1998. The Concept of Resilience: Retrospect and Prospect, *Environment and Development Economics*, 3, 235-239.

_____ . 1999. Contemporary Research in Ecological Economics: Five Outstanding Issues, *International Journal of Ecology and Environmental Sciences*, 25, 143-154.

_____ . 2000. Quantifying the Transient Response of Ecological-Economic Systems to Perturbations, *Environmental Impact Assessment Review*, 20, 125-133.

_____ . 2001. On the Optimal Number of Paddocks in Short Duration Grazing, *Journal of Economic Research*, 6, 159-169.

Batabyal, A.A., and Beladi, H. 2002. On the Choice of the Optimal Temporal Control in Renewable Resource Management, *Stochastic Environmental Research and Risk Assessment*, 16, 325-332.

Bork, E.W., West, N.E., and Walker, J.W. 1998. Cover Components on Long-Term Seasonal Sheep Grazing Treatments in Three-Tip Sagebrush Steppe, *Journal of Range Management*, 51, 293-300.

Holechek, J.L., Pieper, R.D., and Herbel, C.H. 2001. *Range Management*, 4th edition. Prentice Hall, Upper Saddle River, NJ.

Holling, C.S. 1973. Resilience and Stability of Ecological Systems, *Annual Review of Ecology and Systematics*, 4, 1-23.

_____ . 1996. Engineering Resilience versus Ecological Resilience, in P.C. Schulze, (Ed.), *Engineering Within Ecological Constraints*. National Academy Press, Washington, DC.

Holling, C.S., Schindler, D.W., Walker, B.W., and Roughgarden, J. 1995. Biodiversity in the Functioning of Ecosystems: An Ecological Synthesis, in C. Perrings, K.G. Maler, C. Folke, C.S. Holling, and B.O. Jansson, (Eds.), *Biodiversity Loss*. Cambridge University Press, Cambridge, UK.

Ives, A.R. 1995. Measuring Resilience in Stochastic Systems, *Ecological Monographs*, 65, 217-233.

Krebs, C.J. 1985. *Ecology*, 3rd edition. Harper and Row, New York, NY.

Neubert, M.G., and Caswell, H. 1997. Alternatives to Resilience for Measuring the Responses of Ecological Systems to Perturbations, *Ecology*, 78, 653-665.

Perevolotsky, A., Ne'eman, G., Yonatan, R., and Henkin, Z. 2001. Resilience of Prickly Burnet to Management in East Mediterranean Rangelands, *Journal of Range Management*, 54, 561-566.

Perrings, C. 1998. Resilience in the Dynamics of Economy-Environment Systems, *Environmental and Resource Economics*, 11, 503-520.

Pimm, S.L. 1984. The Complexity and Stability of Ecosystems, *Nature*, 307, 321-326.

Ross, S.M. 2002. *Probability Models for Computer Science*. Harcourt Academic Press, San Diego, CA.

Savory, A., and Butterfield, J. 1999. *Holistic Management*. Island Press, Washington, DC.

Chapter 10

HERDING IN RANCH OPERATIONS: AN ANALYSIS OF A SEARCH PROBLEM

ABSTRACT

Herding is an integral part of ranch operations in contemporary times. An aspect of herding that has received scant attention in the literature concerns the pros and cons of searching for livestock animals that have wandered away from a pasture and are hence missing from this pasture. Under what circumstances should a rancher search for these missing animals? We provide the first theoretical analysis of this question. Specifically, we construct and examine a formal model that provides us with conditions under which it makes sense to search for these missing animals and conditions under which it does not.

Keywords: Herding, Optimal Search, Ranch Operation

1. INTRODUCTION

To students of the use and the management of rangelands, herding is salient in two ways. From the perspective of a range manager, herding can significantly improve the distribution of grazing animals in mountain riparian and other ranges. Consequently, a range manager will often want to manage rangelands by judiciously herding grazing animals (Holechek *et al.* (2001, p. 303)). This notwithstanding, our interest in herding stems from the fact that it is an integral part of ranching operations in contemporary times in the United States.

Modern ranch operations in the United States are of various types. Some consist of lands that are entirely privately owned. Others are part-private and part-public in the sense that they consist of privately owned lands and lands that are owned by the Bureau of Land Management (BLM) and/or the Forest Service (FS).[1] In all these ranch operations, two kinds of herding activities are performed routinely. The first kind of herding involves making sure that the animals stay in a particular pasture or allotment and that they don't wander off

[1] An example of a ranch operation that consists of privately owned land, BLM land, and FS land is the Saval Ranch in Nevada. For more on this ranch operation, see Torell *et al.* (1986).

elsewhere. The second kind of herding involves physically moving animals from one pasture to another pasture or alternately to a BLM allotment.

When a pasture or allotment is large in area and unfenced, and it is not uncommon for this to be the case, ranchers often cannot account for all their animals after a period of grazing. They may lose animals permanently as a result of deaths. This can happen when animals that have wandered to the distant corners of a pasture/allotment or off the pasture/allotment consume poisonous plants, or when they are exposed to cold weather for prolonged periods, and when they are killed by predators such as coyotes.[2] In contrast, a rancher may lose some of his animals temporarily. This happens when the rancher's animals are either on a distant part of the pasture/allotment in question or in a neighboring pasture/allotment and they have to be herded back to some desired location.

In any event, when animals are missing, a rancher has to decide whether to invest costly resources and search for them. Depending on the odds of finding the missing animals, a decision to search can have significant financial ramifications for our rancher (see Torell *et al.* (1986)). Consequently, under *what circumstances* should the rancher search for his missing animals? To the best of our knowledge, this important question has not been analyzed previously in the literature. As such, in what follows, we provide the *first* theoretical analysis of this question. In the next section, we first construct and then conduct a detailed analysis of a formal model that provides us with conditions under which it makes sense for a rancher to search for his missing animals and conditions under which it does not. The final section concludes and offers suggestions for future research.

2. The Search Problem

2.1. Preliminaries

Denote the number of missing animals in a specific pasture by N. After a particular grazing period and when the rancher's animals are being counted, N is a deterministic variable. However, looked at over time and certainly from a planning perspective, N is a random variable. To this end, we suppose that N is Poisson distributed with mean θ.[3] If our rancher chooses to search for his missing animals either on the pasture under study or in neighboring pastures/allotments then each animal will independently be found after a length of time that is exponentially distributed with rate β.[4] For each animal that is found by our rancher, he receives a benefit of $\$B$. As well, this rancher incurs a cost of $\$C$ per unit of search time. Finally, we suppose that our rancher spends $t \in 5_+$ units of time searching for his missing animals before calling off the search.

---

2 For more on sheep deaths as a result of predation by coyotes, see Arthur (1981) and Knowlton *et al.* (1999).

3 The Poisson process has been used frequently in the natural resource economics and in the mathematical ecology literatures. For a more detailed corroboration of this claim, see Uhler and Bradley (1970), Pielou (1977), Arrow and Chang (1980), Batabyal (2001), and Batabyal *et al.* (2001).

4 The exponential distribution has been used previously in the ecological and in the natural resource economics literatures to model lifetimes and other variables related to time. For more on this, see Batabyal (2000) and Batabyal and Beladi (2001). Finally, for more on the Poisson and the exponential distribution functions, see Ross (2000).

Our task now is threefold. First, we need to use the information in the previous paragraph to compute our rancher's expected benefit from searching for his missing animals. Next, we would like to determine the optimal t. In other words, given the previous paragraph's description of benefits and costs, what is the optimal length of time for which our rancher should be searching for his missing animals? Finally, note that a policy of searching for a fixed (but arbitrary) time period is a static search policy. Consequently, we would like to know whether our rancher can do any better by following a dynamic search policy. In this context, a dynamic search policy is one which allows the decision to stop searching at any time t to depend on the number of missing animals found by time t. We now address each of these three questions sequentially.

2.2. Analysis

2.2.1 The Expected Benefit Function

We first compute our rancher's expected benefit from searching for his missing animals. To this end, note that because each missing animal will be found independently after searching for a time that is exponentially distributed, we have

$$\textit{Probability of finding each missing animal} = 1-\exp(-\beta t). \tag{1}$$

Using equation (1), we can tell that the total number of missing animals found is a Poisson distributed random variable with mean $\theta\{1-\exp(-\beta t)\}$. Now using this information along with equation (1), we conclude that our rancher's expected benefit from searching for his missing animals is

$$E[\textit{benefit}] = B\theta\{1-\exp(-\beta t)\} - Ct, \tag{2}$$

where $E[\bullet]$ is the expectation operator.

Equation (2) contains the rancher's objective function that we're after. We see that our rancher's expected benefit function up to time t is given by the mean number of missing animals found $\{\theta\{1-\exp(-\beta t)\}\}$ multiplied by the benefit per missing animal found $\{B\}$ less the cost per unit of search time $\{C\}$ multiplied by the relevant time period $\{t\}$.

2.2.2. The Optimal t

The second part of our threefold task is to determine the optimal value of t, i.e., the optimal length of time for which our rancher should be searching for his missing animals. This optimal t can be determined by maximizing equation (2) over t. As such, our rancher solves

$$\max_{\{t\}}[B\theta\{1-\exp(-\beta t)\} - Ct]. \tag{3}$$

The optimal t, t^*, is given by the first order necessary condition for a maximum.[5] Performing the necessary computations, we get

$$t^* = \frac{1}{\beta} \log_e \left\{ \frac{B\theta\beta}{C} \right\}, \tag{4}$$

provided that $B\theta\beta > C$. On the other hand if $B\theta\beta \le C$ then $t^* = 0$ is optimal.

The above finding has an interesting implication. Specifically, we learn that it is optimal to search for missing animals for a *positive time period* provided that the benefit per animal found multiplied by the mean number of missing animals and the rate β exceeds the cost per unit of search time, i.e., if $B\theta\beta > C$ holds. On the other hand, when $B\theta\beta \le C$ holds, because the optimal t, t^*, is a non-negative real number, it is optimal to let the missing animals go and *not search at all.* This interesting finding is consistent with reality. Available evidence[6] indicates that, on occasion, ranchers do not search for their missing animals but instead write them off as a cost of doing business.

2.2.3. The Merits of a Dynamic Search Policy

We have just discussed the pros and cons of pursuing a static search policy. Our last task now is to determine whether our rancher does better with a dynamic search policy. Recall that a dynamic search policy is one which allows the decision to stop searching at any time t to depend on the number of missing animals found by time t.

Now observe that the number of missing animals not found by any time t is independent of the number found because each of the Poisson number of missing animals will independently either be found with probability $\{1-\exp(-\beta t)\}$ or not be found with probability $\{\exp(-\beta t)\}$. Consequently, there is no added benefit whatsoever in letting the decision on whether to stop searching at time t depend on the number of missing animals already found. In other words, in the setting of this chapter, static and dynamic search policies are *equivalent* and our rancher is as well served with a simple (static) search policy as he is with a more complex (dynamic) search policy.

3. CONCLUSIONS

In a ranching operation with herding, under what circumstances should the rancher search for his missing animals? We analyzed this hitherto unstudied question by studying it in a search-theoretic framework. First, we derived our rancher's expected benefit function from searching with a so-called static search policy. Next, given that our rancher is following a static search policy, we determined the optimal length of the search period. We showed that there are circumstances in which it is optimal to not search at all. Finally, we showed that in the context of this chapter, our rancher gains nothing by following a dynamic search policy in

5 The reader can easily verify that the second order sufficiency condition is satisfied.

6 Personal communication dated 12/5/01 from E. Bruce Godfrey, Extension Economist at Utah State University, Logan, UT 84322-3530, USA.

which the decision to stop searching at any time t depends on the number of missing animals found by time t.

The analysis of this chapter can be extended in a number of different directions. In what follows, we suggest two possible extensions. First, we used the Poisson and the exponential distribution functions to conduct our analysis. Although these are commonly used functional forms in the natural resource economics and the mathematical ecology literatures, it would be useful to know whether the results of this chapter hold when the underlying analysis is conducted with general functional forms. Second, when such general functional forms are used, it would be worthwhile to determine the conditions under which the equivalence result of this chapter holds. Studies of the rancher's search problem that incorporate these aspects of the subject into the analysis will provide additional insights into herding in ranch operations in contemporary times.

REFERENCES

Arrow, K.J., and Chang, S.S.L. 1980. Optimal Pricing, Use, and Exploration of Uncertain Natural Resource Stocks, in P.T. Liu, (Ed.), *Dynamic Optimization and Mathematical Economics*. Plenum Press, New York, NY.

Arthur, L.M. 1981. Coyote Control: The Public Response, *Journal of Range Management*, 34, 14-15.

Batabyal, A.A. 2000. An Analysis of Persistence, Resilience, and the Conservation of Keystone Species, *International Journal of Ecology and Environmental Sciences*, 26, 1-10.

_____ . 2001. On Flood Occurrence and the Provision of Safe Drinking Water in Developing Countries, *Applied Economics Letters*, 8, 751-754.

Batabyal, A.A., and Beladi, H. 2001. Aspects of the Optimal Management of Aquatic Ecological-Economic Systems, *European Journal of Operational Research*, 132, 561-568.

Batabyal, A.A., Biswas, B., and Godfrey, E.B. 2001. The Stocking Rate Versus Time in Range Management: A Review of Recent Theoretical Developments, *Applied Economics Letters*, 8, 555-559.

Holechek, J.L., Pieper, R.D., and Herbel, C.H. 2001. *Range Management*, 4th edition. Prentice Hall, Upper Saddle River, NJ.

Knowlton, F.F., Gese, E.M., and Jaeger, M.M. 1999. Coyote Depradation Control: An Interface Between Biology and Management, *Journal of Range Management*, 52, 398-412.

Pielou, E.C. 1977. *Mathematical Ecology*. Wiley, New York, NY.

Ross, S.M. 2000. *Introduction to Probability Models*, 7th edition. Harcourt Academic Press, San Diego, CA.

Torell, L.A., Godfrey, E.B., and Nielsen, D.B. 1986. Forage Utilization Cost Differentials in a Ranch Operation: A Case Study, *Journal of Range Management*, 39, 34-39.

Uhler, R.S., and Bradley, P.G. 1970. A Stochastic Model for Determining the Economic Prospects of Petroleum Exploration Over Large Regions, *Journal of the American Statistical Association*, 65, 623-630.

Chapter 11

SOME THEORETICAL ASPECTS OF RANGE CONDITION ASSESSMENT IN THE PRESENCE OF A THRESHOLD

ABSTRACT

Ecologists have advocated the use of the concept of thresholds in range condition (or state) assessments. Despite this advocacy, there is virtually no theoretical research on range condition assessment in the presence of thresholds. As such, this chapter has three objectives. First, we use a Markov chain theoretic framework to characterize a rangeland whose spatial and temporal behavior is subject to the presence of a threshold. Next, we use this framework to demonstrate the formal relationship between the desirable and the undesirable sets of states of this rangeland. Finally, we compute the expected amount of time this rangeland spends in the desirable and in the undesirable sets of states. These two expectations can be interpreted as a binary measure of the condition of the rangeland under study.

Keywords: Markov Chain, Range Condition, Threshold, Uncertainty

1. INTRODUCTION

Generally speaking, there are two approaches to assessing the condition (or state) of a rangeland.[1] The first approach is based on the seminal work of Dyksterhuis (1949). In this approach, seasonal fluctuations in the forage composition of a particular rangeland are compared to an ecological benchmark that denotes the vegetation climax and this benchmark is assumed to have undergone the same seasonal fluctuations. As is now well known, there are numerous problems with this approach.[2] What is relevant for our purpose is that this approach is *not* well suited to woodland and forested rangelands (Friedel (1991)).

---

[1] In the rest of this chapter, we shall use the terms "condition" and "state" and "range" and "rangeland" interchangeably. The reader should note that some researchers such as Friedel (1991) have used the term "domain" to refer to the condition or the state of a rangeland.

[2] For more on this, see Wilson (1984), Smith (1989), Friedel (1991) and Laycock (1991).

The second approach follows from the work of Humphrey (1949). In this approach (also called the site potential approach), the productive potential of a particular rangeland for a specific use is assessed. As Friedel (1991) and others have pointed out, a difficulty with this approach is that depending on *use*, a single state of the rangeland under study can have a number of potentials. Consequently, one cannot determine, for instance, the extent to which a particular rangeland is degraded by simply inspecting the condition rating of this rangeland. Why not? This is because the condition rating itself is a function of the potential use.

In addition to these problems, neither of these two approaches adequately accounts for the fact that the spatial and the temporal behavior of most rangelands is subject to the presence of *thresholds*. In the context of rangelands, particularly arid and semi-arid rangelands, a threshold has two characteristics. First, it denotes the boundary between two states of a rangeland in space and in time. Second, the move from the pre-threshold state to the post-threshold state is typically reversible only at great cost. Put differently, there is a hysteresis effect at work here. For example, a grassland state will often change into a shrub invaded grassland state when a certain threshold is crossed and it is generally very costly for a range manager to move the rangeland back from the shrub invaded grassland state to the grassland state.

This discussion tells us that thresholds play an important role in determining the condition of a range. Moreover, the concept of a threshold is entirely consistent with the prominent state-and-transition model of range behavior proposed by Westoby *et al.* (1989). Due to these reasons, Friedel (1991) and Laycock (1991) have suggested that researchers pay careful attention to thresholds in range condition assessments. However, because there is virtually *no* theoretical research on range condition assessment in the presence of thresholds, Friedel (1991) has rightly pointed out that substantial theoretical research is needed before the threshold concept can be applied widely in range condition assessments.

Given the paucity of theoretical research in this area,[3] there are many outstanding questions that confront researchers. For instance, if we think of the states of a rangeland before it crosses some threshold as the desirable set of states and the set of states after it crosses this threshold as the undesirable set of states, then what is the formal relationship between the desirable and the undesirable sets of states of this rangeland? Second, what is the expected amount of time that this rangeland spends in the desirable and in the undesirable sets of states? The purpose of this chapter is to answer these two questions.

The rest of this chapter is organized as follows: Section 2 describes the Markov chain[4] theoretic framework that we use in this chapter to model a dynamic and stochastic rangeland. This framework is consistent with the state-and-transition model of Westoby *et al.* (1989). Section 3 uses this theoretical framework to demonstrate the formal relationship between the desirable and the undesirable sets of states of our rangeland. Section 4 computes the expected amount of time that our rangeland spends in the desirable and in the undesirable sets of states. These expectations are important because they can be interpreted as a binary measure of the condition of our rangeland. Finally, section 5 concludes and offers suggestions for future research on range condition assessment in the presence of thresholds.

---

[3] We were able to identify only one theoretical paper, Batabyal (2003), that analyzes issues related to the presence of thresholds. However, the purpose of the Batabyal (2003) paper is to derive measures of the persistence of ecological-economic systems (such as rangelands) whose spatial and temporal behavior is subject to the presence of thresholds. As such, this paper and our chapter have dissimilar objectives.

[4] For more on Markov chains, see Kulkarni (1995, pp. 64-160) and Ross (2000, pp. 163-240).

2. The Theoretical Framework

As explained but not formalized in Westoby *et al.* (1989), rangeland dynamics can be usefully described in terms of a discrete number of conditions and a set of transitions between these conditions. Further, Perrings (1998, p. 517) has recently suggested that researchers use "a Markov approach..." to study the behavior of jointly determined ecological-economic systems such as rangelands. Consistent with these observations, we use a Markov chain theoretic framework to model a dynamic and stochastic rangeland whose spatial and temporal behavior is subject to the presence of a threshold. This framework builds on and generalizes the analysis contained in Batabyal and Godfrey (2002, also chapter 7).

We suppose that our rangeland can exist in a finite number of possible states and we denote this finite number by $n \in \mathbb{1}$. Environmental factors such as droughts, fires, and rainfall, and managerial actions such as grazing, seeding, and the alteration of the stocking rate, result in our rangeland making transitions from one state to another. Further, for any two states i and $j, i \neq j$, the probability of making a transition from state i to state $j, i, j = 1,...,n$ is given by $p_{ij}(e,m)$, where e and m represent environmental factors and managerial actions, respectively. In what follows, for notational simplicity, we shall suppress this dependence of the transition probabilities on environmental factors and managerial actions but the reader should be clear on the point that environmental factors and managerial actions collectively determine the state transitions of the rangeland under study.

In practice, our rangeland will typically exist in a number of short-lived states under alternate combinations of fire, grazing, seasons, and local variability. For the most part, these states do not preclude a move to other short-lived states. However, certain combinations of fire, grazing, seasons, and local variability may push the rangeland "into a new state that is not readily reversed; e.g., an eroded soil surface can develop following heavy grazing coupled with torrential rain, in which case the [rangeland] has crossed a threshold" (Friedel, 1991, p. 422).

When our rangeland crosses a threshold, it is in an altogether different state and the set of states of the rangeland after it has crossed this threshold is qualitatively different and less desirable than the set of states in which it existed before it crossed the threshold. One problem in applying the threshold concept to range condition assessments is that we generally do not know where the threshold lies. This notwithstanding, given the objectives of this chapter, we can certainly partition the state space of our rangeland into a desirable and an undesirable set of states where the threshold marks the boundary between these two sets of states.[5] Let us denote the desirable and the undesirable set of states by D and U respectively. Our task now is to use this Markov chain theoretic framework and demonstrate the formal relationship between D and U.

[5] The reader will note that we are analyzing the case of a single threshold. For some discussion of multiple thresholds, see section 5.

3. The Relationship between the Desirable and the Undesirable Sets of States

Let π_i, $i = 1,...,n$ denote the steady state probabilities of our rangeland. Now recall that π_i represents the fraction of all transitions that are coming from state i. Further, the transition probability p_{ij} denotes the fraction of transitions from state i that are into state j. Combining these two pieces of information, we conclude that the rate at which transitions occur from state i to state j is

$$\textit{rate of transitions } i \rightarrow j = \pi_i p_{ij} . \tag{1}$$

Now, suppose $j \in U$. Then, the rate at which transitions occur from a desirable state to state j is given by

$$\textit{rate of transitions } D \rightarrow j = \sum_{i \in D} \pi_i p_{ij} . \tag{2}$$

We can now determine the rate at which transitions take place from the desirable to the undesirable set of states. This rate can also be thought of as the rate at which our rangeland crosses the threshold. Mathematically, this rate is

$$\textit{range at which rangeland crosses threshold} = \sum_{j \in U} \sum_{i \in D} \pi_i p_{ij} . \tag{3}$$

We are now in a position to demonstrate the formal relationship between the desirable and the undesirable states of our rangeland. To this end, let $\overline{D}(\overline{U})$ denote the expected amount of time our rangeland spends in the desirable (undesirable) set of states. From the properties of Markov chains — see Ross (2000, pp. 178-186) — it follows that, on average, our rangeland crosses the threshold every $\overline{D} + \overline{U}$ time units. This tells us that the rate at which this rangeland crosses the threshold is

$$\textit{range at which rangeland crosses threshold} = \frac{1}{\overline{D} + \overline{U}} . \tag{4}$$

Equating the left-hand-sides (LHS) of equations (3) and (4) we get an equation linking the transition and the steady state probabilities of our rangeland to $\overline{D}$ and $\overline{U}$. That equation is

$$\sum_{j \in U} \sum_{i \in D} \pi_i p_{ij} = \frac{1}{\overline{D} + \overline{U}} . \tag{5}$$

Equation (5) gives us the mathematical relationship that we are after. In words, this equation tells us that as a result of the ongoing impacts of environmental factors and managerial actions, our rangeland crosses the threshold under consideration at a rate equaling the product of the steady state and the transition probabilities, summed over all possible states that this rangeland can exist in. We now compute the expected amount of time that our rangeland spends in the desirable and in the undesirable sets of states.

4. Expected Time Spent in the Desirable and in the Undesirable Sets of States

To compute $\overline{D}$ and $\overline{U}$ explicitly, we need a second equation linking these two expectations. To obtain this second equation, note two things. First, the fraction of time that our rangeland is in the desirable set of states is $\sum_{i\in D}\pi_i$. Second, we can also say that, on average, the fraction of time that the rangeland is in the desirable set of states is $\overline{D}/(\overline{D}+\overline{U})$. Putting these two pieces of information together, we get our second equation. That equation is

$$\sum_{i\in D}\pi_i = \frac{\overline{D}}{\overline{D}+\overline{U}}. \tag{6}$$

We are now in a position to provide closed-form expressions for the two expectations of interest, namely, $\overline{D}$ and $\overline{U}$. Solving equations (5) and (6) simultaneously and then simplifying, we get

$$\overline{D} = \frac{\sum_{i\in D}\pi_i}{\sum_{j\in U}\sum_{i\in D}\pi_i p_{ij}},\ \overline{U} = \frac{\sum_{i\in U}\pi_i}{\sum_{j\in U}\sum_{i\in D}\pi_i p_{ij}}. \tag{7}$$

Equation (7) gives us closed-form expressions for the expected amount of time our rangeland spends in the desirable and in the undesirable sets of states. We see that the denominator of both expressions is identical. Inspecting the numerator of these two expectations we see that the expected amount of time spent in either the desirable or the undesirable states depends significantly on the *number* of states in D and in U.

Equation (7) also provides us with a binary measure of the condition of our rangeland and hence this equation is useful in range condition assessments. Interpreting this measure, we would say that our rangeland's condition is good if $\overline{D}$ is a relatively high number and $\overline{U}$ is a relatively low number. More generally, the higher (lower) the value of $\overline{D}(\overline{U})$ relative to $\overline{U}(\overline{D})$ the better (worse) the condition of our rangeland. To apply this binary measure of range condition in a particular practical setting, it will be necessary to first collect adequate data. Once this has been done, extant statistical methods can be used to estimate the transition and the steady state probabilities of the underlying Markov chain. A discussion of precisely how this can be done is beyond the scope of this chapter but the interested reader is referred to Billard and Meshkani (1995) and to Stolpe (1995) for a more detailed discussion of the relevant methodology.

Recall that environmental factors and managerial actions together affect the transition probabilities of our rangeland. In turn, these transition probabilities influence $\overline{D}$ and $\overline{U}$ and hence our binary measure of range condition. Consequently, it is certainly possible to influence the condition of a range by taking appropriate managerial actions. Moreover, from a range management perspective, it makes sense to view the two expressions in equation (7) as parts of appropriately defined objective functions. In fact, there are two ways of viewing this "ecological-economic" approach to the range management problem. In the first view, the task of a range manager is to choose actions m to *maximize* the mean time that our rangeland spends in the desirable set of states $\overline{D}$ at cost, say, $c_D(m)$. In the second view, the range manager's task is to select actions m to *minimize* the average time the rangeland spends in the undesirable set of states at cost, say, $c_U(m)$.

One key difference between the above two views of the management problem arises from the two cost functions $c_D(m)$ and $c_U(m)$. Because of the presence of a threshold, in general, we expect $c_U(m) > c_D(m)$ to hold. In words, the cost of bringing this rangeland back from the undesirable to the desirable set of states *exceeds* the cost of ensuring that this rangeland does not cross the threshold (stays in the desirable set of states). This observation is entirely consistent with the work of Friedel (1991) and Laycock (1991). Moreover, from a theoretical standpoint, this observation suggests that the appropriate optimization problem to be focusing on is the above described maximization problem. We now conclude and suggest ways in which the research of this chapter might be extended.

5. CONCLUSIONS

In this chapter, we used a Markov chain theoretic framework to shed theoretical light on two hitherto unanswered questions concerning range condition assessment in the presence of a threshold. First, with regard to the formal relationship between the desirable and the undesirable sets of states, we showed that our rangeland crosses the threshold at a rate equaling the product of the steady state and the transition probabilities, summed over all possible states that the rangeland can exist in. Second, we computed the expected amount of time that our rangeland spends in the desirable and in the undesirable sets of states and then we pointed out that these two expectations can be viewed as a binary measure of the condition of our rangeland.

The analysis contained in this chapter can be extended in a number of different directions. In what follows, we suggest two possible extensions. First, we analyzed the case of a single threshold. However, in practice, a particular rangeland may be affected by more than one threshold. Therefore, it would be useful to redo the analysis of this chapter to account for scenarios in which the state space of the rangeland under study is partitioned into more than two sets of states. Second, although the Markov chain theoretic framework of this chapter is consistent with the state-and-transition model of Westoby *et al.* (1989), this framework is somewhat limiting because there is no role for history here. In other words, in a Markov chain theoretic framework, what will happen to a rangeland tomorrow depends only on what is happening today and not on what happened yesterday. As such, it would be useful to model rangelands with more general stochastic processes in which there is a role for history. An examination of these aspects of the problem will allow richer analyses of the ways in which environmental factors and managerial actions affect range behavior and condition.

REFERENCES

Batabyal, A.A. 2003. The Persistence of Ecological-Economic Systems: Alternate Measures and their Properties. *Annals of Regional Science*, 37, 323-336.

Batabyal, A.A., and Godfrey, E.B. 2002. Rangeland Management Under Uncertainty: A Conceptual Approach, *Journal of Range Management*, 55, 12-15.

Billard, L., and Meshkani, M.R. 1995. Estimation of a Stationary Markov Chain, *Journal of the American Statistical Association*, 90, 307-315.

Dyksterhuis, E.J. 1949. Condition and Management of Rangeland Based on Quantitative Ecology, *Journal of Range Management*, 2, 104-115.

Friedel, M.H. 1991. Range Condition Assessment and the Concept of Thresholds: A Viewpoint, *Journal of Range Management*, 44, 422-426.

Humphrey, R.R. 1949. Field Comments on the Range Condition Method of Forage Survey, *Journal of Range Management*, 2, 1-10.

Kulkarni, V.G. 1995. *Modeling and Analysis of Stochastic Systems*. Chapman and Hall, London, UK.

Laycock, W.A. 1991. Stable States and Thresholds of Range Condition on North American Rangelands: A Viewpoint, *Journal of Range Management*, 44, 427-433.

Perrings, C. 1998. Resilience in the Dynamics of Economy-Environment Systems, *Environmental and Resource Economics*, 11, 503-520.

Ross, S.M. 2000. *Introduction to Probability Models*, 7th edition. Harcourt Academic Press, San Diego, CA.

Smith, E.L. 1989. Range Condition and Secondary Succession: A Critique, in W.K. Lauenroth and W.A. Laycock, (Eds.), *Secondary Succession and the Evaluation of Rangeland Condition*. Westview Press, Boulder, CO.

Stolpe, M. 1995. *Technology and the Dynamics of Specialization in Open Economies*. Kieler Studien Number 271, Mohr, Tubingen, Germany.

Westoby, M., Walker, B., and Noy-Meir, I. 1989. Opportunistic Management for Rangelands Not at Equilibrium, *Journal of Range Management*, 42, 265-273.

Wilson, A.D. 1984. Points of Reference in the Assessment of Change in Vegetation and Land Condition, *Australian Rangeland Journal*, 6, 69-74.

Chapter 12

A DYNAMIC AND STOCHASTIC PERSPECTIVE ON THE ROLE OF TIME IN RANGE MANAGEMENT

with B. Biswas and E. B. Godfrey

ABSTRACT

This chapter uses a new ecological-economic approach to analyze the role of time in range management in a dynamic and stochastic setting. We first construct a theoretical model of a parcel of rangeland in which time restrictions are used to manage the land. We then show how the dynamic and the stochastic properties of this rangeland can be used to construct two managerial objectives that are ecologically and economically meaningful. Finally, using these two objectives, we discuss an approach to range management in which the manager has two interrelated goals. This manager maximizes the profits from range operations and (s)he also takes steps to move the rangeland away from the least desirable state of existence.

Keywords: Ecological-Economic System, Semi-Markov Process, Time Restriction, Uncertainty

1. INTRODUCTION

Time restrictions have been used to regulate the activities of productive units in the western world at least since the Fairs and Markets Act of 1448 in England. Since then, time restrictions have been used to limit the number of hours during which shops can remain open and to control other kinds of trading activities. Consider the case of natural resources in contemporary times. Weninger and Strand (1998) have pointed out that recreational and commercial hunters for most game are subject to seasonal restrictions. Moreover, such hunters are commonly required to hunt during daylight hours. Batabyal and Beladi (2002b) have observed that in virtually every state in the USA, sport fishing seasons exist for a whole host of fish species. Commercial fisheries in Canada, the USA, and in western Europe are subject to a variety of time restrictions.[1] This tells us that today, the use of time restrictions for natural resource management is widespread.

Although time restrictions have been and are used to manage natural resources, they have infrequently been used to manage rangelands.[2] In an early contribution, Hormay and Evanko (1958) advocated the need to rest plants from grazing during certain periods of time. However, Hormay and Evanko's ideas did not receive widespread recognition until the emergence of Allan Savory's *Holistic Resource Management* in 1988. In this book and in subsequent work (see Savory and Butterfield (1999)) Savory has forcefully argued in favor of adopting a time based approach to range management. Specifically, he has contended that from a management perspective, it is important to note that overgrazing bears "little relationship to the number of animals but rather to the *time* plants [are] exposed to the animals" (Savory and Butterfield, 1999, p. 46, emphasis in original).[3]

Although this view appears to be gaining currency in the range management profession, very little is known about the *theoretical* properties of time based range management regimes. This is largely because there are *no* analyses that provide an integrated ecological-economic perspective on how a time based management regime affects the ecology and the economics of rangelands. Economists have studied the subject of range management in considerable detail.[4] However, these studies have rarely paid any attention to the *ecological* aspects of the management problem or to the use of *time* as a control variable. In general, these studies have focused on the number of animals that maximize an economic objective function such as a rancher's profit function.

Similarly, although there are many ecological studies of range management,[5] these studies have rarely analyzed *time* restrictions and the *economic* effects of such restrictions. As indicated previously, Hormay and Evanko (1958), Savory (1988) and Savory and Butterfield (1998) have discussed some of the pros and cons of a time based approach to range management. However, because this discussion is largely descriptive, very little is known about the *theoretical* properties of time based management regimes. In addition to this, the three studies just mentioned pay scant attention to the economic aspects of range management.

Given this state of affairs, the present chapter has three goals. First, we build a theoretical model of a rangeland in which time restrictions are used to manage the rangeland. This model accounts for the ecological *and* the economic aspects of the range management problem. Second, we construct two managerial objectives that are ecologically and economically meaningful. These two managerial objectives incorporate in them specific probabilities that are derived from the dynamic and the stochastic properties of the rangeland. These probabilities can be given distinct ecological interpretations. An implication of viewing the range management problem in this way is that unlike most of the previous literature on this

---

1 For additional details on this subject, the reader should consult Karpoff (1987), Hartwick and Olewiler (1998, pp. 152-175), Batabyal (2001a, 2001b, also chapter 3), Batabyal and Beladi (2002a, 2002b), and Xu and Batabyal (2002).

2 In the context of rangelands, a time restriction refers to the length of time during which a rangeland is closed to grazing.

3 Allan Savory's views on grazing have been variously described as short-duration grazing, the Savory grazing method, and as time-controlled grazing. For more on the practical applications of Savory's views, see Savory and Parsons (1980), Savory (1983), and Holechek *et al.* (2001, pp. 269-277)

4 Examples include Huffaker *et al.* (1989), Torrell *et al.* (1991), and McCluskey and Rausser (1999).

5 See Graetz (1986), Gutman *et al.* (1999), and Rehman *et al.* (1999).

subject, our managerial objectives are explicitly ecological-economic in nature. Finally, using these two objectives, we discuss an approach to range management in which the manager has two interrelated goals. This manager maximizes the profits from range operations and (s)he also takes steps to move the rangeland away from the least desirable state of existence.

To the best of our knowledge, this is the *first* essay to provide an integrated approach to range management in a way that accounts for the ecological and the economic aspects of the underlying problem. We stress that the specific purpose of our chapter is to provide a *theoretical* perspective on the role of time in range management. We provide citations to the practical range management literature and we also mention the nexuses between our chapter and practical range management. Nevertheless, the reader should note that our objective in this chapter is *not* to conduct an investigation into the practical aspects of time based range management regimes.

Previous studies of resource management that are related to this chapter are the ones by Perrings and Walker (1995) and by Batabyal and Beladi (2002b). Like this chapter, Perrings and Walker (1995) provide an ecological-economic analysis of the range management problem. However, there are two key differences between this chapter and the Perrings and Walker (1995) paper. First, the decision variable in Perrings and Walker (1995) is the level of offtake and in our chapter it is the time restriction, i.e., the length of time during which grazing is terminated on the rangeland. Second, although Perrings and Walker (1995) are concerned about the resilience[6] of rangelands, they do not explicitly account for this concept in their optimization problems. In contrast, we do. Batabyal and Beladi (2002b) also provide an ecological-economic analysis of time restrictions in natural resource management. However, the specific methods of analysis and the goals of their paper and our chapter are very different. In particular, unlike the goals of this chapter, Batabyal and Beladi (2002b) are primarily interested in determining the likelihood of resource collapse when a resource is managed with time restrictions.

The rest of this chapter is organized as follows: Section 2 presents a semi-Markov model of a rangeland and constructs two objectives for the manger of our rangeland. Section 3 uses these objectives and discusses two optimization problems that a range manager might solve. Section 4 concludes and offers suggestions for future empirical and theoretical research on range management over time and under uncertainty.

2. Time Restrictions in a Dynamic and Stochastic Rangeland

2.1. Preliminaries

Recently, Perrings (1998) has suggested that researchers use a Markovian approach to study jointly determined ecological-economic systems. Further, a Markovian approach nicely captures the essential elements of the "state-and-transition" model of range ecology.[7] This

---

6 Resilience refers to "the amount of disturbance that can be sustained [by a rangeland] before a change in system control or structure occurs." (Holling *et al.*, 1995, p. 50).

7 For more on the "state-and-transition" model and the policy implications of this model see Westoby *et al.* (1989) and Batabyal and Godfrey (2002, also chapter 7).

chapter is the first to use the theory of semi-Markov processes[8] to model a rangeland. Consider a stochastic process with states 0,1,2,3,..., that is now in state i, $i \geq 0$, and that has the following two properties: First, the probability that it will next enter state j, $j \geq 0$, is given by the transition probability P_{ij}. Second, given that the next state to be entered is j, the time until the transition from state i to state j is a random variable with a general distribution function $F_{ij}(\cdot)$ where $F_{ij}^{c}(\cdot) = 1 - F_{ij}(\cdot)$. Now let $y(t)$ denote the state of the process at time t. Then $\{y(t) : t \geq 0\}$ is a semi-Markov process.

In other words, a semi-Markov process is a Markov chain with one significant difference. Whereas a Markov chain spends one unit of time in a state before making a transition to some other state, a semi-Markov process stays in a particular state for a random amount of time before making a transition to some other state. Let K_i denote the distribution function of the time that $\{y(t) : t \geq 0\}$ pends in state i before making a transition to some state and let β_i enote the expectation of this time in state i. Finally, let T_{ii} be the time between successive transitions into state i and let β_{ii} be the expectation of T_{ii}. That is, $\beta_{ii} = E[T_{ii}]$. We are now in a position to discuss the attributes of the rangeland that is the subject of this chapter.

2.2. The Dynamic and Stochastic Rangeland

Consider a dynamic and stochastic rangeland that is privately or publically held and whose *condition* can be in any one of three possible states.[9] In the language of range science, at a specific point in time, this rangeland can exist in any one of three possible *condition classes*.[10] Further, the kind of rangeland we have in mind constitutes — in the words of Savory and Butterfield (1999, pp. 30-34) — a brittle environment.[11]

State 1 is the healthiest state of the rangeland. This means that in this state total forage is plentiful, the quality of this forage is high, and hence the rangeland is open for grazing. State 2 is an intermediate state. In this state, forage quality and quantity are both lower than in state 1 but the rangeland is not endangered in either an ecological or an economic sense. Consequently, the rangeland is still open for grazing. However, the manager now monitors

---

[8] Lucid accounts of semi-Markov processes can be found in Medhi (1994, pp. 313-339) and in Ross (1996, pp. 213-218; 2000, pp. 395-397). Our discussion of semi-Markov processes and this section's model are based in part on Ross (1996, pp. 213-218; 2000, pp. 395-397). We stress that our semi-Markov model is considerably more general than either a discrete-time Markov chain model or a continuous-time Markov chain model of a rangeland.

[9] As noted in Holechek *et al.* (2001, p. 184), range "condition refers to the state of health of the range."

[10] Range managers frequently use the term "condition class" to refer to the state of a rangeland. See Stoddart *et al.* (1975, pp. 187-194) and Graetz (1986) for additional details. The justification for limiting the number of states to three is twofold. First, from a management perspective, rangelands are often conceptualized as existing in one of three possible states. As indicated in Box (1978, pp. 19-20), these states are "Excellent or Good," "Fair," and "Poor or Bad." Second, this three state model is simple and it captures the essential features of a dynamic and stochastic rangeland. Consequently, to keep the underlying management issues transparent, we have decided to analyze this three state model. For a practical application of this condition class classification scheme to prairie ranges in the central Great Plain of the United States, see Parker (1969).

[11] An example of such an environment would be either the sagebrush grassland or the salt desert shrubland of the American west. See Holechek *et al.* (2001, pp. 95-100) for more details.

the condition of the rangeland more carefully than in state 1. State 3 is the state in which the rangeland is endangered or least healthy. In this state, forage quality and quantity are low and the rangeland is severely degraded. Consequently, if the manager determines that the rangeland is in state 3, then no further grazing is permitted. In other words, a time restriction is now in place. The reader will note that we have envisioned the condition or health of a range in terms of the quality and the quantity of the total forage available for consumption. However, it is also possible to think of the condition of a range in terms of the population and the diversity of forage plants.

Let us now formalize the above remarks. As a result of ongoing grazing, ecological/environmental factors (droughts, lack of plant nutrients, unusually low soil moisture), and human induced factors (fires), our rangeland stays in state 1 for a mean length of time β_1 and then makes a transition either to state 2 with transition probability P_{12}, or to state 3 with transition probability P_{13}. When the rangeland is in state 2, once again because of the previously mentioned reasons, this rangeland will stay in state 2 for a mean length of time β_2 and then move to state 3 with transition probability P_{23}. When in state 3, grazing on this rangeland is terminated. As a result of the termination of grazing, the rangeland vegetation gradually recovers. It is important to note that the rate of recovery depends in part on the *extent* of rangeland degradation in state 3. What this means for our purpose is that the length of the recovery period — rest period in the words of Savory and Butterfield (1999, pp. 195-215) — or the length of time during which grazing is not permitted (the time restriction) is itself a *random* variable. Denote the mean length of the time restriction by β_3.[12]

The imposition of this time restriction does not guarantee that the rangeland will get back to the most desirable state 1. Rare events[13] may interact with the time restriction in a way that results in the rangeland recovering only to the intermediate state 2. To account for these features of the problem, we suppose that as a result of the time restriction, the rangeland returns either to state 1 with transition probability P_{31}, or to state 2 with transition probability P_{32}. Our task now is to use these dynamic and stochastic attributes of this rangeland and construct two objectives that our manager might optimize. However, before we do this, it is necessary to first say a few words about the applicability of this chapter's methodology to the determination of the fallow period in slash-and-burn agriculture.

Slash-and-burn agriculture (also called swidden agriculture and shifting cultivation) is a common agricultural system in many developing countries.[14] In this system, forest or brush land is cleared with the slash-and-burn method — this releases nutrients held in plant tissues — and the soil is prepared with a dibble stick or a hoe. There is typically little use of irrigation systems or fertilizers and human labor is the single most important factor of production. The cleared land is generally multi-cropped and the cropping period is short. This cropping period is followed by a fallow period. As noted in Gleave (1996), Hofstad (1997), and Coomes *et al.* (2000), the length of the fallow period is an important choice variable in

[12] In one practical time based range management regime in New Mexico, the grazing period was 5 days long and the recovery/rest period appears to have been about 28 days long. For more on this see Fowler and Gray (1986) and Holechek *et al.* (2001, pp. 269-271).

[13] See Perrings and Walker (1995, pp. 192-195).

[14] For more on slash-and-burn agriculture in developing countries, see Swinkels *et al.* (1997), Ekeleme *et al.* (2000), Li *et al.* (2000), Udaeyo *et al.* (2001), and the citations therein.

slash-and-burn agriculture. In particular, for slash-and-burn agriculture to be viable over any reasonable time horizon, the length of the fallow period must be selected optimally. Now, in the context of our chapter's semi-Markov model, if we think of state 3 as the fallow state and β_3 as the mean length of the fallow period, then it is possible to use methods similar to those employed in this chapter to choose the length of the fallow period to optimize, for instance, a farmer's profits from slash-and-burn agriculture. We now return to our range management problem.

2.3. Two Managerial Objectives

2.3.1. The First Objective: Asymptotic Resilience Weighted Profit Function

To derive the first managerial criterion, it will be necessary to compute the steady state probabilities for our three state rangeland. Formally, we are interested in computing $P_i = \lim_{t\to\infty} Prob\{y(t) = i / y(0) = j\}$ for any state j and for states $i = 1,2,3$. In words, given that our rangeland is in state j at time $t = 0$, we want to compute the limiting probability, as time approaches infinity, that the rangeland will be in state i. To perform this computation, let us denote the limiting probabilities of the embedded Markov chain of our rangeland[15] by π_i, $i = 1,2,3$. Now it is well known — see equation 7.23 in Ross (2000, p. 396) — that these limiting probabilities satisfy

$$\pi_j = \sum_{i=1}^{i=3} \pi_i P_{ij}, \ \sum_{j=1}^{j=3} \pi_j = 1. \tag{1}$$

Consequently, using the transition probabilities of the rangeland and equation (1), we can calculate the required limiting probabilities. These are

$$\pi_1 = \frac{P_{31}}{1 + P_{31} + P_{12}P_{31} + P_{32}}, \ \pi_2 = \frac{P_{12}P_{31} + P_{32}}{1 + P_{31} + P_{12}P_{31} + P_{32}}, \ \pi_3 = \frac{1}{1 + P_{31} + P_{12}P_{31} + P_{32}}. \tag{2}$$

To determine the steady state probabilities (the $P_i's$) of the rangeland, we now use equation 7.24 in Ross (2000, p. 396). This equation tells us that the $P_i's$ satisfy

$$P_i = \frac{\pi_i \beta_i}{\sum_{j=1}^{j=3} \pi_j \beta_j}. \tag{3}$$

Equations (2) and (3) together give us the steady state probabilities that we are after. We get

---

[15] For additional details on the embedded Markov chain of a semi-Markov process, see Medhi (1994) and Ross (1996; 2000).

$$P_1 = \frac{P_{31}\beta_1}{P_{31}\beta_1 + (P_{12}P_{31} + P_{32})\beta_2 + \beta_3}, \tag{4}$$

and

$$P_2 = \frac{(P_{12}P_{31} + P_{32})\beta_2}{P_{31}\beta_1 + (P_{12}P_{31} + P_{32})\beta_2 + \beta_3},\ P_3 = \frac{\beta_3}{P_{31}\beta_1 + (P_{12}P_{31} + P_{32})\beta_2 + \beta_3} \tag{5}$$

Equations (4) and (5) show us exactly how these three steady state probabilities depend on the time restriction β_3. Note that this time restriction is applied *only* if the manager determines that our rangeland is in state 3. Therefore, β_3 does not have any direct effect on either the probability of making a transition from state 1 to 3, P_{13}, or on the probability of making a transition from state 2 to 3, P_{23}. The purpose of β_3 is to affect the probability of making a transition to either state 1 or 2 *from state 3*, i.e., the transition probabilities P_{31} and P_{32}.[16]

In the context of this chapter's ecological-economic approach to the range management problem, these steady state probabilities have a distinct ecological meaning. As discussed in Krebs (1985, p. 587), Batabyal (1998a, 1998b, 1999, 2000), Perrings (1998), and Batabyal and Beladi (1999), these probabilities measure the asymptotic *resilience* of the rangeland in each of the three states. As indicated in footnote 10, resilience is an ecological stability property and it refers to "the amount of disturbance that can be sustained [by a rangeland] before a change in system control or structure occurs." (Holling *et al.*, 1995, p. 50). This means that we can think of the resilience of a rangeland as a long run measure of its well-being. Now, if we rank the three states from this well-being perspective, then it should be clear that our rangeland's well-being is highest in state 1 because forage quality and quantity are plentiful in this state and the rangeland vegetation is not degraded. From a well-being perspective, state 2 is a middle-of-the-road state because forage quality and quantity are at an intermediate level. Finally, the rangeland is least well off in state 3 because in this state the rangeland is endangered. In the words of Perrings (1998), state 1 is a "desirable" state and state 3 is an "undesirable" state.

Recall that the range manager terminates grazing on the rangeland if and only if the rangeland is determined to be in state 3. Further, the mean length of this time restriction is β_3. Now, range operations result in revenues and in costs to our manager. To this end, let us denote the revenue and the cost functions (on which more later) in state i, $i = 1,2,3$, by $R_i(\beta_i, \beta_3, w_i)$ and $C_i(\beta_i, \beta_3, a_i)$ respectively. In other words, the revenue from range operations in state i depends on the mean time spent in state i, β_i, the mean length of the time restriction β_3, and the total animal weight gain in state i, w_i. Similarly, the cost in the *ith* state is a function of the mean time spent in state i, β_i, the mean length of the time restriction β_3, and the various activities — such as range condition monitoring — associated

[16] For more on this, see the discussion in the second paragraph after equation (9) and inspect the objective function described in equations (10) and (13).

with successful range management. We proxy these activities in state i by the variable a_i. We are now in a position to state our range manager's objective function. This objective function is the expected profit from range operations. Mathematically, we have[17]

$$Objective\ Function\ (i) = \sum_{i=1}^{i=3} P_i[R_i(\beta_i, \beta_3, w_i) - C_i(\beta_i, \beta_3, a_i)], \tag{6}$$

where the $P_i's$ are given by equations (4) and (5).

Note that the steady state probabilities $P_i's$ in equation (6) denote both the long run proportion of time that our rangeland is in each of the three possible states and the asymptotic resilience of our rangeland in each of these three states. Consequently, these probability weights can be thought of as ecological correction factors to an economic objective function. This is the manner in which the ecology of the rangeland enters our manager's objective function.

In brittle environments of the sort that we are analyzing in this chapter, it is very important to choose the time restriction or recovery period carefully. It is clear that if the time restriction is too short, then our rangeland will not have had enough time to recover to either state 1 or 2 from state 3. However, this does *not* mean that the manager should err on the side of caution and, *ceteris paribus*, make the time restriction long. As Savory and Butterfield (1999, pp. 206-209) have pointed out, a recovery period that is much longer than the amount of time it takes for damaged plants to rebuild their root system is likely to be detrimental to the health of a brittle rangeland.

Moving to the economic side of the range management problem, we have already noted that range operations give rise to revenues and to costs.[18] To comprehend this clearly, first consider the revenue aspect. A well managed rangeland will be able to provide the manager with a flow of revenues in the different states. In state i, these revenues depend on β_i, β_3, and on the animal weight gain w_i. On the cost side, it is necessary to monitor the condition of the rangeland on an ongoing basis, and personnel involved in the various tasks associated with management have to be paid. We proxy these activities in state i with the variable a_i. Consequently, in deciding the length of the time restriction, in addition to the ecology of the rangeland, our manager must also pay attention to both the revenues and the costs from the time restriction. As such, the objective that our manager focuses on is the profit from range operations. When this profit is weighted by the steady state probabilities that denote the resilience of our rangeland, we get expected profit or asymptotic resilience weighted profit — shown in equation (6) — as our managerial objective function. This completes our discussion of the first managerial objective function.

2.3.2. The Second Objective: Transient Resilience in the Conditional Profit Function

The second managerial objective also involves profit maximization, but now the focus of the manager is a little different. As in section 2.3.1, once again we shall take a long run view

---

[17] It is understood that in state 3, β_3 enters the revenue and the cost functions only once.

[18] For more on this see Fowler and Gray (1986), and Holechek *et al.* (2001, chapters 8 and 9).

of the rangeland. Specifically, suppose that at time t the rangeland is in the undesirable state 3. By choosing the time restriction β_3, the manager can affect not only the profits from range operations in the different states but also the state into which the rangeland will next make a transition. Ideally, the manager would like this next state to be the healthiest state, i.e., state 1. To this end, if we let $D(t)$ be the state entered at the first transition after time t, then we can determine the long run conditional probability that the next state visited after t is 1, given that at t the rangeland is in state 3 and that the mean length of the manager's time restriction is β_3. In other words, ideally, we would like to compute $\lim_{t\to\infty} Prob\{D(t)=1/y(t)=3\}$ and then use this probability to construct our manager's objective function. Note that unlike the steady state probabilities that we computed in the previous sub-section, the probability that we now seek is a steady state *conditional* probability.

We are now in a position to compute this probability. Elementary probability theory tells us that

$$\lim_{t\to\infty} Prob\{D(t)=1/y(t)=3\} = \lim_{t\to\infty} \frac{Prob\{D(t)=1, y(t)=3\}}{Prob\{y(t)=3\}}. \tag{7}$$

The joint probability in the numerator of the right-hand-side (RHS) of this equation can be simplified with the aid of Theorem 4.8.4 in Ross (1996, p. 217). The probability in the denominator of the RHS of equation (7) is simply P_3, the steady state probability (see equation (5)) of finding the rangeland in state 3. With these simplifications, we get

$$\lim_{t\to\infty} Prob\{D(t)=1/y(t)=3\} = P_{31} \bullet \frac{\int_0^\infty F_{31}^c(w)dw}{P_3\beta_{33}}. \tag{8}$$

Proposition 4.8.1 in Ross (1996, p. 214) can be used to further simplify the denominator on the RHS of equation (8). This gives

$$\lim_{t\to\infty} Prob\{D(t)=1/y(t)=3\} = P_{31} \bullet \frac{\int_0^\infty F_{31}^c(w)dw}{\beta_3}. \tag{9}$$

The RHS of equation (9) is the product of two terms. The first term is the probability of making a transition from state 3 to 1. The second term is the ratio of the integral of the tail distribution of the amount of time the rangeland spends in state 3 before making a transition to state 1 to the mean length of the manager's time restriction in state 3. Following Perrings (1998), we are now in a position to give an ecological interpretation to this first term. This term is the transient or the short run resilience of the rangeland in the least desirable state 3.

Let us now use the steady state conditional probability in equation (9) to construct our manager's objective function. As in the previous sub-section, our range manager's basic goal is to choose the mean length of the time restriction β_3 to maximize the profits from range operations. However, this time we shall focus on an alternate version of this profit maximization problem. Instead of maximizing the expected or the asymptotic resilience weighted profit from range operations, our manager now solves a *conditional* profit maximization problem in which the profits obtained in state 1 are conditional on the rangeland moving from state 3 to 1. In other words, using the notation of section 2.3.1, our manager is interested in $[R_1(\beta_1, \beta_3, w_1) - C_1(\beta_1, \beta_3, a_1)]$, the profits from range operations in state 1. However, state 1 arises only if the manager is able to choose the time restriction β_3 to move the rangeland from state 3 to 1. Given that the rangeland is currently in state 3, the conditional probability of moving to state 1 is given by equation (9). Consequently, our range manager's objective function now is a conditional profit function and this function is obtained by multiplying $[R_1(\beta_1, \beta_3, w_1) - C_1(\beta_1, \beta_3, a_1)]$ by the probability in equation (9). Mathematically, we have

$$\begin{aligned} &Objective\ Function\ (ii) = \lim_{t \to \infty} Prob\{D(t) = 1 / y(t) = 3\} \bullet \\ &[R_1(\beta_1, \beta_3, w_1) - C_1(\beta_1, \beta_3, a_1)]. \end{aligned} \tag{10}$$

In equation (10), β_1, β_3, w_1, and a_1 are as in the previous sub-section. However, note that because P_{31}, the transient resilience of our rangeland in state 3 is an argument of the stationary conditional probability in equation (9), the range manager's conditional profit function is itself a function of the rangeland's transient resilience in state 3. This is the way in which the ecology of the rangeland enters our manager's objective function.

The economic side of the management problem is a little different from what we had in the previous sub-section. The manager's objective function now is a profit function that is conditional on the rangeland moving from state 3 to 1. In determining the length of the time restriction, in addition to the ecology of the rangeland, our manager will now pay attention to the state 1 revenues and costs that arise probabilistically from the optimal choice of the time restriction. This completes our discussion of the second managerial objective and the associated managerial optimization problem. We now analyze these optimization problems and then draw inferences for range management.

3. Optimal Range Management with Ecological-Economic Criteria[19]

3.1. Maximizing the Asymptotic Resilience Weighted Profit Function

Recall from the discussion in section 2.3.1 that the first problem faced by our manager involves the maximization of an asymptotic resilience weighted profit function subject to a non-negative time restriction. Formally, our manager solves (see equation (6))

---

[19] Our focus in this chapter is on the role of time in range management. This is why we are focusing exclusively on the optimal choice of β_3. This does not mean that the other arguments of the profit function are irrelevant.

$$\max_{\{\beta_3\}} \sum_{i=1}^{i=3} P_i[R_i(\beta_i, \beta_3, w_i) - C_i(\beta_i, \beta_3, a_i)], \tag{11}$$

subject to $\beta_3 \geq 0$.

Now, without loss of generality, suppose that the solution to problem (11) yields an interior maximum. Then, omitting the complementary slackness condition, the Kuhn-Tucker condition for a maximum is

$$K[P_{31}\beta_1 \frac{\partial R_1}{\partial \beta_3} + (P_{12}P_{31} + P_{32})\beta_2 \frac{\partial R_2}{\partial \beta_3} + \beta_3 \frac{\partial R_3}{\partial \beta_3} + \beta_3] - P_{31}\beta_1 R_1 - (P_{12}P_{31} + P_{32})\beta_2 R_2 - \beta_3 R_3 =$$
$$K[P_{31}\beta_1 \frac{\partial C_1}{\partial \beta_3} + (P_{12}P_{31} + P_{32})\beta_2 \frac{\partial C_2}{\partial \beta_3} + \beta_3 \frac{\partial C_3}{\partial \beta_3} + C_3] - P_{31}\beta_1 C_1 - (P_{12}P_{31} + P_{32})\beta_2 C_2 - \beta_3 C_3, \tag{12}$$

where $K = P_{31}\beta_1 + (P_{12}P_{31} + P_{32})\beta_2 + \beta_3$.

The optimal time restriction, say β_3^*, solves equation (12). In words, this equation tells us that in choosing the time restriction optimally, the manager will balance ecological and economic considerations. Specifically, β_3^* will be chosen so that the marginal revenue from the time restriction (the LHS) equals its marginal cost (the RHS).

If the recovery period is chosen in this way, then we can be fairly sure that the rangeland will be healthy in the long run. From an ecological perspective, this means that the resilience of the rangeland in the relatively desirable states (1 and 2) will be high and its resilience in the undesirable state 3 will be low. In economic terms, this means that operations on this rangeland will provide our manager with a stream of profits in the long run.

3.2. Maximizing the Conditional Profit Function with Transient Resilience

Recall from section 2.3.2 that in this version of the management problem, the manager's objective is to choose the time restriction so that the conditional profit function in equation (10) is maximized. Formally (see equations (9) and (10)), our manager now solves

$$\max_{\{\beta_3\}} P_{31} \bullet \frac{\int_0^{\infty} F_{31}^c(w)dw}{\beta_3}[R_1(\beta_1, \beta_3, w_1) - C_1(\beta_1, \beta_3, a_1)], \tag{13}$$

subject to $\beta_3 \geq 0$.

As stated, this maximization problem is unwieldy. Consequently, to illustrate our approach, we shall make a distributional assumption about the amount of time that the rangeland spends in state 3 before making a transition to state 1. Specifically, we suppose that this time is exponentially distributed. Then, integrating the tail distribution function for an exponentially distributed random variable — see Jeffrey (1995, p. 248) — and then evaluating this integral between the upper and the lower limits, we get

$$\int_0^\infty F_{31}^c(w)dw = \frac{1}{\delta}, \tag{14}$$

where $\delta > 0$ is the parameter of the exponential distribution function. Using equation (14), our manager's maximization problem becomes

$$\max_{\{\beta_3\}} \frac{P_{31}}{\delta\beta_3}[R_1(\beta_1,\beta_3,w_1) - C_1(\beta_1,\beta_3,a_1)] \ , \tag{15}$$

subject to $\beta_3 \geq 0$. Now, as in the previous sub-section, suppose that the solution to problem (15) yields an interior maximum. Then, omitting the complementary slackness condition, the Kuhn-Tucker condition for a maximum is

$$\beta_3 \frac{\partial R_1}{\partial \beta_3} - R_1 = \beta_3 \frac{\partial C_1}{\partial \beta_3} - C_1. \tag{16}$$

The optimal time restriction, β_3^*, solves equation (16). This equation can be thought of as an "ecological-economic" optimality condition. In words, equation (16) tells us that when the manager's goal is to maximize the profits from range operations in state 1 after the rangeland has moved to this state from state 3, (s)he will choose the time restriction so that the long run conditional marginal revenue from the time restriction in state 1 (the LHS) is equal to the long run conditional marginal cost in this state (the RHS).

If β_3^* is chosen in this way, then one can be fairly sure that our rangeland will be sustainable in the long run. It is important to stress that in the context of our chapter, sustainability has a dual meaning. From an ecological standpoint, sustainable means that in the long run, the rangeland will not be resilient in the undesirable state 3. From an economic standpoint, sustainable means that the rangeland will provide our manager with a stream of profits in the long run. We now briefly discuss the salience and the policy implications of the research contained in this chapter.

3.3. Salience and Policy Implications of this Research

Holechek *et al.* (2001, p. 271, emphasis added) have noted that "a lack of *long-term* research [has prevented researchers from] drawing very many definite conclusions about the effectiveness of various [time-controlled] grazing strategies." In this chapter, we have provided a theoretical perspective on time-controlled grazing. Four specific policy conclusions follow from this chapter's research. First, unlike most economics papers on this subject, our chapter shows that successful range management involves paying attention to both the ecology and the economics of the rangeland under consideration. Second, we have shown that by maximizing the *long run* profit functions of this chapter, a range manager will also be enhancing the resilience of the desirable states (states 1 and 2) and lowering the resilience of the undesirable state 3. Third, this chapter has shown how a range manager might set the time restriction optimally in an ecological-economic context. Finally, from a practical perspective, this chapter sheds light on the probabilities that will need to be estimated in order to set up the objective functions described in equations (6) and (10). In addition to this, the maximization exercises of sections 3.1 and 3.2 provide the manager with two different ways of setting the time restriction optimally.

4. Conclusions

We addressed three issues in this chapter that, to the best of our knowledge, have not been addressed previously in the range management literature. First, in section 2, we used the theory of semi-Markov processes to provide an ecological-economic model of a rangeland that is managed with time restrictions. Next, we used the dynamic and the stochastic properties of this rangeland to construct two managerial objective functions that are meaningful from an ecological and an economic standpoint. We stress that because these two objectives are ecological-economic in nature, our modeling of the range management problem is quite different from previous approaches to this problem in the economics literature. Finally, in section 3, we used these two objectives to analyze two range management problems from an ecological-economic perspective. In this perspective, the focus of the manager is on using the time restriction to (i) maximize the profits from range operations and (ii) move the rangeland away from the least desirable state of existence.

The analysis contained in this chapter can be extended in a number of different directions. In what follows, we suggest three avenues for empirical and theoretical research on the subject of range management over time and under uncertainty. First, it would be useful to determine whether extant econometric techniques can be used to estimate the transition probabilities of the rangeland under study. Second, given a specific grazing system such as time-controlled grazing, knowledge of the amount of time a rangeland spends in a particular condition class or state would be helpful in setting up the objective functions discussed in this chapter. Finally, we modeled the rangeland as a three state semi-Markov process. Although this is consistent with the "condition class" view of rangelands that is frequently employed by range scientists, it would still be useful to generalize the theoretical analysis of this chapter to an arbitrary but finite number of states. Studies of range management that incorporate these aspects of the problem into the analysis will provide additional insights into the management of rangelands whose behavior is marked by a considerable amount of uncertainty.

References

Batabyal, A.A. 1998a. On Some Aspects of Ecological Resilience and the Conservation of Species, *Journal of Environmental Management*, 52, 373-378.

_____ . 1998b. The Concept of Resilience: Retrospect and Prospect, *Environment and Development Economics*, 3, 235-239.

_____ . 1999. Species Substitutability, Resilience, and the Optimal Management of Ecological-Economic Systems, *Mathematical and Computer Modelling*, 29, 35-43.

_____ . 2000. Quantifying the Transient Response of Ecological-Economic Systems to Perturbations, *Environmental Impact Assessment Review*, 20, 125-133.

_____ . 2001a. Some Theoretical Aspects of Short Duration Grazing, *International Journal of Ecology and Environmental Sciences*, 27, 215-220.

_____ . 2001b. On the Optimal Number of Paddocks in Short Duration Grazing, *Journal of Economic Research*, 6, 159-169.

Batabyal, A.A., and Beladi, H. 1999. The Stability of Stochastic Systems: The Case of Persistence and Resilience, *Mathematical and Computer Modelling*, 30, 27-34.

_____ . 2002a. On the Choice of the Optimal Temporal Control in Renewable Resource Management, *Stochastic Environmental Research and Risk Assessment*, 16, 325-332.

_____ . 2002b. Time Restrictions in Natural Resource Management: A Dynamic and Stochastic Analysis. Forthcoming, *European Journal of Operational Research*.

Batabyal, A.A., and Godfrey, E.B. 2002. Rangeland Management Under Uncertainty: A Conceptual Approach, *Journal of Range Management*, 55, 12-15.

Box, T.W. 1978. The Arid Lands Revisited: 100 Years Since John Wesley Powell. *57th Annual Faculty Honor Lecture*, Utah State University.

Coomes, O.T., Grimard, F., and Burt, G.J. 2000. Tropical Forests and Shifting Cultivation: Secondary Forest Fallow Dynamics Among Traditional Farmers of the Peruvian Amazon, *Ecological Economics*, 32, 109-124.

Ekeleme, F., Akobundu, I.O., Isichei, A.O., and Chikoye, D. 2000. Influence of Fallow Type and Land-Use Intensity on Weed Seed Rain in a Forest/Savanna Transition Zone, *Weed Science*, 48, 604-612.

Fowler, J.M., and Gray, J.R. 1986. Economic Impacts of Grazing Systems During Drought and Non-Drought Years on Cattle and Sheep Ranches in New Mexico, *New Mexico Agricultural Experiment Station Bulletin 725*, New Mexico State University.

Gleave, M.B. 1996. The Length of the Fallow Period in Tropical Fallow Farming Systems: A Discussion With Evidence from Sierra Leone, *Geographical Journal*, 162, 14-24.

Graetz, R.D. 1986. A Comparative Study of Sheep Grazing in a Semi-Arid Saltbush Pasture in Two Condition Classes, *Australian Rangeland Journal*, 8, 46-56.

Gutman, M., Holzer, Z., Baram, H., Noy-Meir, I., and Seligman, N.G. 1999. Heavy Stocking and Early-Season Deferment of Grazing on Mediterranean-Type Grassland, *Journal of Range Management*, 52, 590-599.

Hartwick, J.M., and Olewiler, N.D. 1998. *The Economics of Natural Resource Use*, 2nd edition. Addison-Wesley, Reading, MA.

Hofstad, O. 1997. Degradation Processes in Tanzanian Woodlands, *Forum for Development Studies*, 0, 95-115.

Holechek, J.L., Pieper, R.D., and Herbel, C.H. 2001. *Range Management*, 4th edition. Prentice Hall, Upper Saddle River, NJ.

Holling, C.S. Schindler, D.W., Walker, B., and Roughgarden, J. 1995. Biodiversity in the Functioning of Ecosystems: An Ecological Synthesis, in C. Perrings, K.G. Maler, C. Folke, C.S. Holling, and B.O. Jansson, (Eds.), *Biodiversity Loss*. Cambridge University Press, Cambridge, UK.

Hormay, A.L., and Evanko, A.B. 1958. Rest-Rotation Grazing: A Management System for Bunchgrass Grazing. *California Forest and Range Experiment Station Miscellaneous Paper*, United States Department of Agriculture.

Huffaker, R.G., Wilen, J.E., and Gardner, B.D. 1989. Multiple Use Benefits on Public Rangelands: An Incentive-Based Fee System, *American Journal of Agricultural Economics*, 71, 670-678.

Jeffrey, A. 1995. *Handbook of Mathematical Formulas and Integrals*. Academic Press, San Diego, CA.

Karpoff, J.M. 1987. Suboptimal Controls in Common Resource Management: The Case of the Fishery, *Journal of Political Economy*, 95, 179-194.

Krebs, C.J. 1985. *Ecology*, 3rd edition. Harper and Row, New York, NY.

Li, F., Zhao, S., and Geballe, G.T. 2000. Water Use Patterns and Agronomic Performance for Some Cropping Systems With and Without Fallow Crops in a Semi-Arid Environment of Northwest China, *Agriculture, Ecosystems, and Environment*, 79, 129-142.

McCluskey, J.J., and Rausser, G.C. 1999. Federal Grazing Reform and Avoidable Risk, *Journal of Agricultural and Resource Economics*, 24, 140-154.

Medhi, J. 1994. *Stochastic Processes*, 2nd edition. Wiley, New Delhi, India.

Parker, K.G. 1969. The Nature and Use of Utah Range, *Utah State University Extension Circular # 359*, Utah State University.

Perrings, C. 1998. Resilience in the Dynamics of Economy-Environment Systems, *Environmental and Resource Economics*, 11, 503-520.

Perrings, C., and Walker, B. 1995. Biodiversity Loss and the Economics of Discontinuous Change in Semiarid Rangelands, in C. Perrings, K.G. Maler, C. Folke, C.S. Holling, and B.O. Jansson, (Eds.), *Biodiversity Loss*. Cambridge University Press, Cambridge, UK.

Rehman, A. Mackintosh, J.B., Warren, B.E., and Lindsay, D.R. 1999. Revegetated Saline Pastures as a Forage Reserve for Sheep: 1. Effects of Season and Grazing on Morphology and Nutritive Value of Saltbush, *Rangeland Journal*, 21, 3-12.

Ross, S.M. 1996. *Stochastic Processes*, 2nd edition. Wiley, New York, NY.

_____ . 2000. *Introduction to Probability Models*, 7th edition. Academic Press, San Diego, CA.

Savory, A. 1983. The Savory Grazing Method or Holistic Resource Management, *Rangelands*, 5, 155-159.

_____ . 1988. *Holistic Resource Management*. Island Press, Washington, DC.

Savory, A., and Parsons, S.D. 1980. The Savory Grazing Method, *Rangelands*, 2, 234-237.

Savory, A., and Butterfield, J. 1999. *Holistic Management*, 2nd edition. Island Press, Washington, DC.

Stoddart, L.A., Smith, A.D., and Box, T.W. 1975. *Range Management*, 3rd edition. McGraw Hill, New York, NY.

Swinkels, R.A., Franzel, S., Shepherd, K.D., Ohlsson, E., and Ndufa, J.K. 1997. The Economics of Short Rotation Improved Fallows: Evidence From Areas of High Population Density in Western Kenya, *Agricultural Systems*, 55, 99-121.

Torell, L.A., Lyon, K.S., and Godfrey, E.B. 1991. Long Run versus Short Run Planning Horizons and the Rangeland Stocking Rate Decision, *American Journal of Agricultural Economics*, 73, 795-807.

Udaeyo, N.U., Umoh, G.S., and Ekpe, E.O. 2001. Farming Systems in Southeastern Nigeria: Implications for Sustainable Agricultural Production, *Journal of Sustainable Agriculture*, 17, 75-89.

Weninger, Q., and Strand, I.E. 1998. Regulating Production Time to Control Output: An Application to a Harvest Technology, *Unpublished Manuscript*, Utah State University.

Westoby, M., Walker, B., and Noy-Meir, I. 1989. Opportunistic Management for Rangelands Not at Equilibrium, *Journal of Range Management*, 42, 266-274.

Xu, Q., and Batabyal, A.A. 2002. A Theoretical Analysis of Season Length Restrictions in Fisheries Management, *Discrete Dynamics in Nature and Society*, 7, 241-247.

INDEX

A

B

C

D

E

F

G

H

I

J

K

L

M

N

O

T

U

V

W

Z